Study Guide for Psychology to Accompany Neil J. Salkind's

Statistics for People Who (Think They) Hate Statistics

6

EDITION

Sara Miller McCune founded SAGE Publishing in 1965 to support the dissemination of usable knowledge and educate a global community. SAGE publishes more than 1000 journals and over 800 new books each year, spanning a wide range of subject areas. Our growing selection of library products includes archives, data, case studies and video. SAGE remains majority owned by our founder and after her lifetime will become owned by a charitable trust that secures the company's continued independence.

Los Angeles | London | New Delhi | Singapore | Washington DC | Melbourne

Study Guide for Psychology to Accompany Neil J. Salkind's

Statistics for
People Who (Think They)
Hate Statistics

6
EDITION

Neil J. Salkind
The University of Kansas

Prepared by Ryan J. Winter
Florida International University

Los Angeles | London | New Delhi
Singapore | Washington DC | Melbourne

FOR INFORMATION

SAGE Publications, Inc.
2455 Teller Road
Thousand Oaks, California 91320
E-mail: order@sagepub.com

SAGE Publications Ltd.
1 Oliver's Yard
55 City Road
London, EC1Y 1SP
United Kingdom

SAGE Publications India Pvt. Ltd.
B 1/I 1 Mohan Cooperative Industrial Area
Mathura Road, New Delhi 110 044
India

SAGE Publications Asia-Pacific Pte. Ltd.
3 Church Street
#10–04 Samsung Hub
Singapore 049483

Acquisitions Editor: Helen Salmon
Editorial Assistant: Chelsea Neve
Production Editor: Andrew Olson
Typesetter: Hurix Systems Pvt. Ltd.
Proofreader: Jennifer Grubba
Cover Designer: Candice Harman
Marketing Manager: Shari Countryman

Printed in the United States of America

ISBN 978-1-5063-9573-9

This book is printed on acid-free paper.

17 18 19 20 21 10 9 8 7 6 5 4 3 2 1

Contents

General Outline

Following is a general outline showing the sequence of content presented for each chapter.

Chapter Outline

Learning Objectives

Summary/Key Points

Key Terms

True/False Questions

Multiple-Choice Questions

Exercises

Short-Answer/Essay Questions

SPSS Questions

Just for Fun/Challenge Yourself

Answer Key

ADDITIONAL PRACTICE OPPORTUNITIES

Appendix A on page 193 contains *Practice With Real Data!* The data sets for these exercises are available on the main website for this text and Study Guide at **edge.sagepub.com/Salkind6e**. You can find the data sets by clicking on **Resources for the Study Guide for Psychology** on the left-hand navigation.

1

Statistics or Sadistics?
It's Up to You

LEARNING OBJECTIVES

- Understand the purpose and scope of statistics.

- Review (briefly) the history of statistics.

- Get an introduction to descriptive and inferential statistics.

- Review the benefits of taking a statistics course.

- Learn how to use and apply this book.

SUMMARY/KEY POINTS

Introduction to Part I

- Researchers in a very wide variety of fields use statistics to make sense of the large sets of data they collect in studying a great number of interesting problems.
  - Michelle Lampl, a pediatrician and anthropologist, has studied the growth of infants, finding that some infants can grow as much as 1 inch overnight.
  - Sue Kemper, a professor of psychology, has studied the health of nuns, finding that the complexity of the nuns' writing during their early 20s is related to risk for Alzheimer's disease as many as 70 years later.
  - Aletha Huston, a researcher and teacher, has found that children who watch educational programs on television do better in school than those who don't.

- Statistics can be defined as "the science of organizing and analyzing information," making that information easier to understand.

- Statistics are used to make sense of often large and unwieldy sets of data.

- Statistics can be used in any field to answer a very wide variety of research questions and hypotheses.

A Brief History of Statistics

- Far back in human history, collecting information became an important skill.

- Once numbers became part of human language, they began to be attached to outcomes. In the 17th century, the first set of data relating to populations of people was collected.

- Once scientists began to collect data, they needed to develop specific tools to answer specific questions. This led to the development of statistics, which can look at both the frequency of events (like the number of children born with autism in the United States in a specific decade) as well as differences between groups (like whether people who are under a great deal of stress at work are more likely to yell at their spouses at home compared to people who are more relaxed at work).

- In the early 20th century, the simplest test for examining the differences between the averages of two groups was developed. For example, are people who take "cat naps" during their workday more productive than those who do not take cat naps? Or are couples who are similar happier than couples who are dissimilar (the whole "birds of a feather flock together" versus "opposites attract" debate)?

- The development of powerful and relatively inexpensive computers has revolutionized the field of statistics. While individuals can now conduct complex and computationally intensive statistical analyses with their own computers, they can potentially run analyses incorrectly or arrive at incorrect conclusions regarding their results. A researcher may incorrectly claim that there is a statistically significant difference between a group that sleeps too much and a group that sleeps too little on a cognitive task when there are no actual differences between groups.

- Today, researchers from a wide variety of fields use basically the same techniques, or statistical tests, to answer very different questions. This means that learning statistics enables you to conduct quantitative research in almost any field. It also means that you can understand the methods used in journal articles written by social psychologists, school psychologists, environmental psychologists, clinical psychologists, developmental psychologists, and neuropsychologists, as well as dozens of other specialties within the field of psychology.

Statistics: What It Is (and Isn't)

- Statistics describes a set of tools and techniques that is used for describing, organizing, and interpreting information or data. It helps us understand the world around us.

- Descriptive statistics are used to organize and describe the characteristics of a collection of data. The collection is sometimes called a data set or just data. This might involve finding the average scores on an IQ test or the frequency of times children disobey their parents. The participant section of most journal articles include descriptive statistics related to participant gender, ethnicity, and age.

- Inferential statistics are often (but not always) carried out after descriptive statistics. They are used to make inferences from a smaller group of data to a larger one. An example is using results from one kindergarten classroom to infer, or generalize, about a population of a whole kindergarten grade. Or a researcher might infer whether the skills children with Attention Deficit Hyperactivity Disorder learn in a skills education class can transfer to other populations or other settings.

- A sample is a portion or subset of a larger population. Data from samples may be used for description only, or to generalize something about the larger population.

- A population is a full set from which a sample is taken: all the possible cases of interest. Data from a sample can be used to infer properties of a whole population. Of course, the sample should represent the larger population. If a study finds high ratings of confidence in elected officials, but it only uses community members with a high socio-economic status (SES), the results may not apply to those with lower SES.

Why Study Statistics? What Am I Doing in a Statistics Class?

- Having statistical skills puts you at an advantage when applying to graduate school or for a research or academic position.

- If not a required course for your major, a basic statistics course on your transcript sets you apart from other students.

- A statistics course can be an invigorating intellectual challenge.

- Having a knowledge of statistics makes you a better student, as it will enable you to better understand journal articles and books in your field as well as what your professors and colleagues study and discuss. It will also make you a critical consumer of information outside of an educational setting. When logging onto the Internet, reading a newspaper or magazine, or watching the local news, you are often exposed to research results. But just how valid and reliable are those studies? This course will give you valuable insight into the methods used in those research reports, allowing you to determine whether the studies are dependable.

- A basic knowledge of statistics will position you well for further study if you plan to pursue a graduate degree in the social or behavioral sciences or in many other fields.

Tips for Using This Book

- Be confident: Work hard, and you'll do fine.

- Statistics is not as difficult as it's made out to be.

- Don't skip chapters: Work through them in sequence.

- Form a study group.

- Ask your professor questions.

- Do the exercises at the end of each chapter.

- Practice, practice, practice: Besides the exercises, find other opportunities to use what you've learned.

- Look for applications to make the material more real. If available at your university, try to join a research lab. You can get great exposure to research by actually engaging in research with a faculty member or graduate student.

- Browse: Flip through the future material and review chapters.

- Have fun: Enjoy mastering a new field and acing your course.

KEY TERMS

- **Statistics**: A set of tools and techniques that is used for describing, organizing, and interpreting information or data

- **Descriptive statistics**: A set of statistical techniques and tools that is used to organize and describe data

- **Data, Data set**: A set of data points (where one data point = one observation/measurement)

- **Inferential statistics**: A set of statistical techniques and tools that is used to make inferences from a smaller group of data to a larger one

- **Sample**: A subset of the population. A researcher's goal is often to generalize findings from a sample to a population

- **Population**: All the possible subjects or cases of interest

2 Means to an End

Computing and Understanding Averages

LEARNING OBJECTIVES

- Understand averages, or measures of central tendency, one of the key components of descriptive statistics.

- Learn how to calculate the mean, median, and mode.

- Understand the distinction among these three measures of the average.

- Understand that the mean is very sensitive to outliers.

- Understand and apply scales or levels of measurement.

- Understand which measures of the average to use with different types of data.

- Learn how to use SPSS to calculate measures of central tendency.

SUMMARY/KEY POINTS

- Averages, or measures of central tendency, are used to determine the single value that best represents an entire group of scores. Popular measures of the average include the mean, median, and mode.

- The mean consists of the middle point of a set of values, and it is simply the sum of all the values divided by the number of values.

- The median consists of the middle point of a set of cases, and the mode represents the most frequent value in a set of scores.

- Only the mode can be used when determining an average for qualitative, categorical, or nominal data.

- Likewise, the median and mean can only be used with quantitative data.

- The mean can be considered the most precise measure, followed by the median and, finally, by the mode. While the mean is the most precise measure, it is sensitive to extreme scores. The median is not sensitive to extreme scores, so it better represents the center-most value of a data set that does include extreme scores.

- Scales of measurement are key to choosing the correct measure of central tendency to use. There are certain characteristics of data, specifically the scale of measurement, and each level of the scale builds upon the previous. The scale of measurement can be at the nominal, ordinal, interval, or ratio levels. The more precise levels of measurement (for example, the interval level of measurement) contain all the qualities of the scales below them.
  - **Nominal** variables are by class or category, are the least precise, and are mutually exclusive. Gender, ethnicity, and political affiliation can be nominal variables. In a psychology study, a nominal variable may involve asking a participant whether they recall someone telling them they did a poor job on an exam (and thus attacking their self-esteem) or did well on an exam (and thus bolstering their self-esteem). There are two categories here (did a poor job versus did a good job), though you could add other categories (was not told about their performance). **Ordinal** variables are measured in order or rank. The order is noted, but there is no way to tell the amount of difference between each rank. Class rankings, sports rankings, or finishers in a race could all be ordinal variables. In a psychology study, an ordinal variable may involve asking children to rank a series of toys. The researcher can then see if the rankings change depending on whether the child is allowed (or not allowed) to play with the one they

ranked second-favorite. Here, all you know is the rank order. The first and second preferred toy may vie for the top position, with the third and fourth toy not well liked at all. **Interval** variables are frequently used in tests or assessment tools in which there is some underlying continuum that indicates the amount of a higher or lesser value. Intervals, or spaces or points, along a continuum are equal to one another. In a psychology study, an interval variable may involve asking participants to rank their level of fear on a scale of 1 (no fear at all) to 9 (terrified) when shown a video about either spiders (experimental condition) or butterflies (comparison condition). An interval scale could also entail negative numbers depending on the needs of the researcher. The fear of spiders scale, for example, could range from -10 (very fearful) to $+10$ (not at all fearful). An assessment tool at the **ratio** level of measurement means that it has an absolute zero (e.g., weight or length). Ratio is the highest level of scales of measurement. In a psychology study, a ratio scale may involve asking participants to complete a ten-question recall exam about whether a picture of an African-American or Caucasian appeared on a computer screen. Here, participants cannot get fewer than zero questions correct. Measures like age, weight, and height also have a zero point. Although it is *near* impossible to weigh zero pounds, it is *definitely* impossible to weigh negative pounds!

  ○ The easiest way to recall the scale of measurement order (least informative to most informative) is to think about the science of NOIR: Nominal, Ordinal, Interval, and Ratio!

- The following are guides for when to use what scale of measurement (but there can be exceptions): Use the mode when variables are nominal. Use the median when you have extreme scores and you do not want to distort the average. Use the mean when you have data that do not include extreme scores and the variables are not categorical. The nominal level of measurement is the least precise, while the ratio level of measurement is the most precise. The "higher up" you are on the scale of measurement, the more precise, detailed, and informative your data are.

- For a sample statistic, Roman letters are used. For a population parameter, Greek letters are used.

KEY TERMS

- **Average**: The one value that best represents an entire group of scores. This can be the mean, median, or mode.

- **Measures of central tendency**: Another term for *averages*. As in the definition of average, measures of central tendency consist of the mean, median, and mode. These descriptive statistics *describe* the central number in a data set.

- **Mean**: The sum of all the values in a group, divided by the number of values in the group
  ○ The mean is sometimes represented using X-bar, or the letter M, and it is also called the typical, average, or most central score.
  ○ The mean is very sensitive to extreme scores.
  ○ When calculating the mean by hand, computing the "weighted mean" can save time when your data contain multiple instances of different values.
  ○ In a psychology study, a researcher might ask participants to rate the extent to which watching a violent versus nonviolent television show affected their levels of anger (1 = I do not feel angry at all to 9 = I feel a great deal of anger).

- **Median**: The midpoint of a set of scores
  ○ The median is sometimes abbreviated as *Med* or *Mdn*.

- To compute the median, list all the values in order, from highest to lowest or lowest to highest. Next, find the middle-most score. If you have an even number of values, the median is calculated as the mean of the two middle values.
- **Percentile points**: These are used to define the percentage of cases equal to and below a certain point in a distribution or set of scores. A score at the 25th percentile (Q_1) is at or above 25% of the other scores in the distribution. A score at the 50th percentile is often referred to as Q_2 and is the median.
- Because the median focuses on cases, and not the values of those cases, it is much less sensitive to extreme scores, or outliers, than is the mean.
- In a psychology study, a researcher might measure the arrest rates of police officers who respond to calls of domestic violence. Imagine the police department has a "star" officer, whose arrest rates in a one-year span are 95 arrests. Most officers in the study have arrest rates in the 35 to 40 range. The researcher may want to use the median arrest rate here, as the outlier "95 arrests" greatly impact the mean.

- **Mode**: The value that occurs most frequently in a set of data
  - To find the mode, first list all the values in the distribution, listing each value only once. Next, count the number of times that each value occurs. The one that occurs most often is the mode.
  - If a set of values has more than one mode, the distribution is multimodal.
  - A distribution can be multimodal even if it has multiple modes that are very similar but not exactly the same (i.e., 15 of one category and 16 of some other category).
  - In a psychology study with 50 participants, a researcher might want to know how many participants are married, currently dating, or single. If 25 are married, 15 are dating, and 10 are single, the mode would be "married."

- **Skew** (verb): When your data include too many extreme scores, the distribution of scores can become *skewed*, or significantly distorted.

- **Data points**: Individual observations in a set of data

- **Scales of measurement**: Different levels at which outcomes are measured. The four scales of measurement are nominal, ordinal, interval, and ratio. A good way to remember these is the acronym NOIR
  - **Nominal level of measurement**: The level of measurement such that outcomes can only be placed into unranked categories, like guilt (guilty versus not guilty), gender (male versus female), type of vehicle owned (sedan, pick-up truck, van, motorcycle), restaurant drink options (tea, soda, coffee, juice, milk, water)
  - **Ordinal level of measurement**: The level of measurement such that outcomes can be rank ordered. For example, consider drink preference. A child may prefer soda, then juice, then milk, and finally water. Although he may have a clear preference for soda and juice, he may not like milk and water at all. All you know is the order in an ordinal variable, not the degree of preference.
  - **Interval level of measurement**: The level of measurement such that outcomes are based on some underlying continuum that makes it possible to speak about how much greater one performance is than another. Many scaled questions in psychology are based on interval scales. Such scales may ask someone to rate their fear, anxiety, happiness, sadness, preference, confidence, or beliefs.
  - **Ratio level of measurement**: The level of measurement such that outcomes are based on some underlying continuum that also contains a true, or absolute, zero. Although rarer in psychology, ratio scales can show a complete lack of some quality. For example, someone can get zero questions correct on a recall test. Though difficult, they can take zero seconds to respond to the presence of a picture flashed on a computer screen. They cannot take less than zero seconds!

TRUE/FALSE QUESTIONS

1. The mode and median are both averages.

2. The mean is very sensitive to extreme scores.

3. The more precise levels of measurement (for example, the interval level of measurement) contain all the qualities of the scales below them.

4. Ordinal variables have two features. They show order or rank and they have equal distance between points along a scale.

5. If a researcher asked participants to rate how much they like a person who just told them they failed an exam (1 = I like them very much to 9 = I do not like them at all), the researcher used a ratio scale of measurement.

6. A researcher gives participants five minutes to solve a series of riddles. At the end of the five minutes, the researcher counts how many riddles the participant solved correctly. The number of solved riddles uses a ratio scale of measurement.

MULTIPLE-CHOICE QUESTIONS

1. Of the set of values {170, 249, 523, 543, 572, 689, 1,050}, what is 543?

   a. The mean

   b. The median

   c. The mode

   d. The percentile

2. What is the mean of the following set of values: 1,501, 1,736, 1,930, 1,176, 446, 428, 768, and 861?

   a. 1,105.75

   b. 1,018.5

   c. 428

   d. 1,930

   e. 8

3. Your data set contains a variable on region of residence that contains the following possible responses: Northeast, South, Midwest, West, and Pacific Coast. Which of the following measures of central tendency should you use for this variable?

   a. The mean

   b. The median

   c. The mode

   d. The weighted mean

   e. Both a and b

4. This is the level of measurement where outcomes are based on some underlying continuum such that it is possible to speak about how much greater one performance is than another one.

   a. Nominal

   b. Ordinal

   c. Interval

   d. Ratio

5. This is the level of measurement such that outcomes can be placed only into unranked categories.

   a. Nominal

   b. Ordinal

   c. Interval

   d. Ratio

6. This is the level of measurement such that outcomes are based on an underlying continuum that contains a true, or absolute, zero.

   a. Nominal

   b. Ordinal

   c. Interval

   d. Ratio

7. This is the level of measurement such that outcomes can be rank ordered.

   a. Nominal

   b. Ordinal

   c. Interval

   d. Ratio

8. This is the most precise level of measurement.

   a. Nominal

   b. Ordinal

   c. Interval

   d. Ratio

9. This is the least precise level of measurement.

   a. Nominal

   b. Ordinal

   c. Interval

   d. Ratio

10. A researcher is interested in eyewitness memory. She has participants watch a video reenactment of a robbery and then gives them ten True / False questions about the event. She counts the number of correct responses. The responses are BEST assessed using which scale of measurement?

    a. Nominal

    b. Ordinal

    c. Interval

    d. Ratio

11. A researcher wants to look at popular food items in a college cafeteria. He asks students what their favorite food is. Which scale of measurement would he use to describe the most popular foods mentioned by the participants?

    a. Nominal

    b. Ordinal

    c. Interval

    d. Ratio

12. You want to determine whether the race of the defendant has an impact on jury verdicts. You assign participants to watch a trial with either a Hispanic or Caucasian defendant, and you measure on a 1 (not at all) to 7 (very) scale how guilty each participant thinks the defendant is. What is the best DV scale of measurement?

    a. Nominal

    b. Ordinal

    c. Interval

    d. Ratio

13. Let's say you want your DV to be relevant to the real legal world, thus you just look at guilty vs. not guilty options. Which scale of measurement should you use now?

    a. Nominal

    b. Ordinal

    c. Interval

    d. Ratio

14. A researcher wants to gauge student preferences for a new textbook. He gives students three books, and has them rank them in order of preference. Which scale of measurement is the researcher MOST LIKELY using

    a. Nominal

    b. Ordinal

    c. Interval

    d. Ratio

15. Imagine we buy a bag of balloons. When we count them up, we find there are 18 red, 12 blue, 25 orange, 24 purple, and 21 green balloons. What is the mode?

     a. 18

     b. 21

     c. 25

     d. None of the above

16. Last year, a fast food outlet in a beachside city paid 3 kitchen hands $16,000 per year each, 2 supervisors $22,000 each, and the owner $85,000. What is the mean?

     a. $29,500

     b. $19,000

     c. $16,000

     d. $14,500

17. Last year, a fast food outlet in a beachside city paid 3 kitchen hands $16,000 per year each, 2 supervisors $22,000 each, and the owner $85,000. What is the mode?

     a. $29,500

     b. $19,000

     c. $16,000

     d. $14,500

18. Last year, a fast food outlet in a beachside city paid 3 kitchen hands $16,000 per year each, 2 supervisors $22,000 each, and the owner $85,000. What is the median?

     a. $29,500

     b. $19,000

     c. $16,000

     d. $14,500

EXERCISES

1. Calculate the mean for the following set of values: 25, 37, 53, 72, 76.

2. Calculate the median for the same set of values: 25, 37, 53, 72, 76.

3. Calculate the mode for the following set of values: 7, 7, 12, 15, 17, 19, 22, 25, 27, 31, 35, 42, 42, 47, 59.

4. Imagine a researcher wants to look at the intelligence quotient (IQ) levels of a sample of participants. Their IQs are: 89, 92, 103, 104, 108, 121, and 134. What is the mode IQ of the participants? What is the median IQ of the participants? Finally, what is the mean IQ of the participants?

SHORT-ANSWER/ESSAY QUESTIONS

1. Review the following set of values: 12, 24, 37, 42, 55, 62, 72, 77, 246, 592. What would be the best measure of central tendency to use for this set of values? Why?

2. You have just finished conducting a study on college students that contained a large set of questions on demographic information, including about the participants' gender, major, year in college (freshman, sophomore, junior, or senior), and race. What measure of central tendency should be used for these types of variables? Why?

3. In a survey that asked individuals about their favorite styles of music, 37 replied rock, 27 said they preferred pop music, 14 stated they liked classical music the most, and 3 individuals stated that they preferred opera. Based on these data, what would be the mode? Why couldn't the mean or median be used to describe this set of values?

4. You just took the GRE (in preparation for graduate school) and received a total score in the 98th percentile. Is this a good or bad score?

5. Come up with an example of an ordinal level of measurement. Now, what would be an example of an interval level of measurement?

6. A researcher is interested in testing a new drug therapy program for children with Attention Deficit Hyperactivity Disorder (ADHD). He assigns some students to the new drug while others get a placebo. The researcher then measures their ability to pay attention to their teacher on a scale from 1 (shows little improvement) to 9 (shows a great deal of improvement). Which scale of measurement would you use here and why? In addition, which measure of central tendency should you use and why?

SPSS QUESTIONS

1. A psychologist asks 25 participants who are currently dating to rate the likelihood that they will still be in the relationship in twelve months on a scale of 0 (very likely they will not remain in the relationship) to 100 (very likely they will still be in the relationship). Data shows the following 25 scores: 72, 13, 79, 76, 29, 8, 12, 27, 90, 72, 29, 40, 22, 45, 28, 50, 40, 84, 71, 14, 56, 46, 25, 28, 33. Input these scores in SPSS and use SPSS to calculate the mean, median, and mode.

   Now, go through the numbers yourself and calculate these averages by hand. Make sure your calculations match those produced by SPSS.

JUST FOR FUN/CHALLENGE YOURSELF

1. A set of five values was found to have a mean of 54.8. You know that four of the values are 24, 27, 53, and 68. What is the fifth value?

2. It is close to the end of the semester, and your goal is to get at least an A– (an average of 90) in your class. Each of your three exams is worth 25% of your grade, and your final is also worth 25%. You received an 87 on your first exam, an 88 on your second exam, and a 91 on your third exam. What is the minimum score (represented as a whole number) that you need to get on your final in order to get your A–?

TRUE/FALSE QUESTIONS

1. True. The mean, median, and mode are all different averages.

2. True.

3. True.

4. False. Ordinal variables only show order or rank. Interval variables have equal distance between points along a scale.

5. False. Since there is no absolute zero, a ratio scale is not appropriate for this question. An interval scale is best.

6. True. It is possible for participants to solve zero riddles, but they cannot solve fewer than zero riddles.

MULTIPLE-CHOICE QUESTIONS

1. (b) The median

2. (a) 1,105.75

3. (c) The mode

4. (c) Interval

5. (a) Nominal

6. (d) Ratio

7. (b) Ordinal

8. (d) Ratio

9. (a) Nominal

10. (d) Ratio

11. (a) Nominal (He didn't ask them to rank foods, just list them.)

12. (c) Interval

13. (a) Nominal

14. (b) Ordinal (The books are ranked, but book #1 might be the favorite by far.)

15. (d) None of the above (The correct mode is "orange". This is a categorical variable, so look at the categories—red, blue, orange, purple, and green. There are 25 orange balloons, so "orange" is the most frequently occurring score.)

16. (a) $29,500

17. (c) $16,000

18. (b) $19,000 (Rank order the scores. $22,000 and $16,000 are both central numbers, so add them and divide by 2.)

EXERCISES

1. 52.6

2. 53

3. Modes = 7 and 42.

4. Mode = all scores (89, 92, 103, 104, 108, 121, and 134); Median = 104; Mean = 107.29.

SHORT-ANSWER/ESSAY QUESTIONS

1. Based on this set of values, the best measure of central tendency would be the median. First, the mode is only preferred in situations in which your variable is qualitative or categorical (i.e., situations in which the mean or median cannot be computed). Second, the mean should not be used in this situation because you have two very high outliers, 246 and 592.

2. For these types of variables, you can only use the mode. Since these variables are categorical/qualitative, it would be impossible to calculate the mean. Likewise, the median cannot be computed for variables of this nature.

3. Based on these data, the mode would be "rock music," as this is the most common category of response. The mean or median could not be used with this type of variable as it is a qualitative/categorical variable.

4. This score is very good: Only 2% of people who took this exam did better than you did.

5. An ordinal level of measurement is any variable that is categorical (consisting of a number of discrete categories) and that can be ordered, or ranked. Some examples include highest degree earned, social class, and letter grade. An interval level of measurement is any variable that is continuous but does not have an absolute or true zero. Some examples are IQ, height, weight, and grade on an exam (in these examples, it is assumed there is no "true" zero).

6. The effectiveness of the drug is based on an interval scale of measurement (1 = Shows little improvement to 9 = Shows a great deal of improvement). The best measure of central tendency is the mean, which is the most informative measure of center. However, should there be outliers in the resulting data (a few really high or really low scores that do not reflect other scores in the distribution), the researcher should use the median, the second most informative measure of center.

SPSS QUESTIONS

1. Statistics

Var	
Valid	25
Missing	0
Mean	43.5600
Median	40.0000
Mode	28.00a

Multiple modes exist. The smallest value is shown (28). Other modes include 29, 40, and 72

Mean = 43.56, Median = 40

JUST FOR FUN/CHALLENGE YOURSELF

1. $\bar{X} = \dfrac{\sum X}{n} \Rightarrow \dfrac{24+27+53+68+x}{5} = 54.8 \Rightarrow \dfrac{172+x}{5} = 54.8 \Rightarrow 172+x = 274 \Rightarrow x = 102.$

2. $\bar{X} = \dfrac{\sum X}{n} \Rightarrow \dfrac{87+88+91+x}{4} = 90 \Rightarrow \dfrac{266+x}{4} = 90 \Rightarrow 266+x = 360 \Rightarrow x = 94.$

3 Vive la Différence

Understanding Variability

LEARNING OBJECTIVES

- Understand what measures of variability are, how they are used, and how they differ from one another.

- Learn how to calculate the range, standard deviation, and variance by hand.

- Learn how to use SPSS to calculate measures of variability.

- Understand the benefits of unbiased estimates.

SUMMARY/KEY POINTS

- In addition to measures of central tendency, measures of variability make up another important component of descriptive statistics.
  - Variability is a measure of how much each score in a group of scores differs from the mean. Consider two data sets:
    Data Set A: 5, 8, 20, 23, and 44
    Data Set B: 17, 19, 20, 21, and 23
    The median for both data sets is 20. The mean is also 20. There are five modes for each data set. Yet which one is most variable? There is lot of variability in Data Set A, but less variability in Data Set B.

- Measures of variability include the range, standard deviation, variance, and mean deviation.
  - The range is the easiest measure of variability to calculate, but it is the most general. It should never be used alone to reach any conclusions regarding variability. Data Set A above has a range of $44 - 5 = 39$. Data Set B has a range of $23 - 17 = 6$.
  - The standard deviation, the most frequently used measure of variability, is a measure of the average distance from the mean. Since the standard deviation uses the same scale as the original data set, it is often reported in journal articles. For example, in the sentence, "Students were more nervous when told their essay would be reviewed by a full professor ($M = 5.33$, $SD = 1.21$) compared to a teaching assistant ($M = 3.55$, $SD = 0.89$)," the M indicates the mean and SD indicates the standard deviation. The standard deviation might also be represented by the lower case letter s ($M = 5.33$, $s = 1.21$).
  - The variance is the standard deviation squared. It is seldom reported in journal articles and is primarily used in statistical formulas as a practical measure of variability.

- Both the standard deviation and the variance include the term $n - 1$. Estimates in which 1 is subtracted from the sample size are called *unbiased estimates* and are more conservative estimates of population parameters.

- One example of variability used in research is in the work of Nicholas Stapelberg and colleagues. They used measures of variability in heart rate as it related to coronary heart disease and depression. You can also see variability in the amount of time it takes to help someone who is no longer breathing. People with a background in medical training (nurses, doctors, EMTs, or those with CPR training) may be quicker to help than people without such training. Thus, "speed to help" can be very variable!

KEY TERMS

- **Variability**: The amount of spread or dispersion in a set of scores

- **Range**: The highest minus the lowest score. This is a very general estimate of the range.
  - **Exclusive range**: The highest score minus the lowest score
  - **Inclusive range**: The highest score minus the lowest score plus 1

- **Standard deviation**: The average amount of variability in a set of scores, or a measure of the average distance from the mean. This is the most common measure of variability, but it is sensitive to extreme scores.

- **Variance**: The square of the standard deviation. This measure is much more difficult to interpret than the standard deviation, which is why it is used much less often.

- **Mean deviation**: The sum of the absolute value of the deviations from the mean divided by the number of scores. This calculation differs from that of the standard deviation.

- **Unbiased estimate**: An estimate of a population parameter in which 1 is subtracted from n. It is considered to be a more conservative estimate than the biased estimate.
  - Biased estimates can be used if you only wish to describe a sample, while unbiased estimates should be used when making estimates of population parameters.

TRUE/FALSE QUESTIONS

1. Two sets of values that have the same mean must also have the same variability.

2. It is possible for two or more sets of values to have the same standard deviation and variance.

3. Before the standard deviation or variance is calculated, the mean of scores needs to be calculated.

4. Like the mean, the standard deviation is sensitive to extreme scores.

5. It is common for a set of scores to have no variability.

6. The variance is easier to interpret than the standard deviation.

7. If the variance score for a set of numbers is 6.78, the standard deviation is 45.97.

8. A researcher finds the following scores for the amount of money people spent on a car: $9,999, $8,500, $8,750, $9,000, $10,000, $60,000. The researcher would want to use the mean to describe the car price.

9. In a study looking at the racial composition of a graduating high school class and whether the students have been accepted into college, the researcher would report the mean and standard deviation for ethnicity.

10. The mean, median, and mode tell you what the central number in a data set is while the range, variance, and standard deviation tell you how much spread or variance there is in the data set.

11. Researchers should report both the standard deviation and variance when writing their results sections.

MULTIPLE-CHOICE QUESTIONS

1. Which of the following sets of values has the greatest variability?

   a. 2, 2, 3, 3, 4

   b. 7, 7, 8, 9, 9

   c. 2, 3, 5, 7, 8

   d. 1, 4, 7, 9, 11

2. In which of the following sets of values is the mean equal to the standard deviation?

   a. 2, 4, 6, 8, 9

   b. 4, 7, 8, 8, 12

   c. 0, 2, 4, 6, 13

   d. 2, 4, 5, 9, 11

3. Which of the following sets of values has no variability at all?

   a. 10, 20, 30, 40, 50

   b. 2, 4, 6, 8, 10

   c. 0, 0, 0, 0, 5

   d. 8, 8, 8, 8, 8

4. What is the most general measure of variability?

   a. The range

   b. The standard deviation

   c. The variance

   d. The mean deviation

5. You are conducting a survey, and you wish to use a sample of respondents in order to estimate population parameters. Which of the following should you use?

   a. Biased estimates

   b. Unbiased estimates

   c. Both unbiased and biased estimates

6. Which of the following statements is correct?

   a. The range is typically equal to the mean.

   b. The range is equal to the square of the variance.

   c. The variance is equal to the square of the standard deviation.

   d. The standard deviation is equal to the square of the variance.

7. Which measure of variability is reported most by psychologists?

    a. The mode

    b. The standard deviation

    c. The variance

    d. The median

8. "Results indicated that participants were more likely to respond aggressively to the obnoxious researcher when they were not being paid for the study ($M = 6.77$, $SD = 1.22$) than when they were being paid for the study ($M = 4.55$, $SD = 2.12$)." Which of the following is true about this sentence?

    a. The researcher calculated the standard deviation by taking the square root of each mean.

    b. The means are equivalent between the two conditions (paid versus unpaid).

    c. The mean represents the central score in each condition while the standard deviation represent the measure of spread in each condition.

    d. The author of the sentence used the incorrect statistics here. He should have reported the variance.

9. Research on modeling behavior finds that participants often mimic the behaviors of those they see engage in an action. Imagine a researcher has a grocery store patron (a confederate) donate to a food pantry at the register. The researcher then measures whether a real patron behind the confederate similarly donates to the food pantry. The researcher measures whether the real patron donates (yes or no) and, if so, how much they donate. The researcher would provide the mean and standard deviation for which of the following:

    a. Both whether the patron donates as well as how much she donates

    b. Neither whether the patron donates nor how much she donates

    c. Whether she donates only

    d. How much she donates only

EXERCISES

1. Compute the range of the following set of scores: 214, 246, 379, 420.

2. Compute the standard deviation and variance of the following scores: 32, 48, 55, 62, 71.

SHORT-ANSWER/ESSAY QUESTIONS

1. When and why are unbiased estimates preferred over biased estimates?

2. A psychologist asks participants who just watched a scary movie to rate their level of fear on a scale from 1 (not at all fearful) to 7 (extremely fearful). The variance for the scores is 11.55. In writing up the results section of her study, should the psychologist report the variance? Why or why not?

3. A psychologist studying the effectiveness of an alcohol abuse program reports that the average age of the participants was 35 with a standard deviation of 3.5 years. Why would the psychologist want to report both the mean age and the standard deviation?

4. Published studies usually provide the demographic characteristics of the sample used in the study, with most researchers providing descriptive statistics for gender, ethnicity, and age in the participant section. Should the researcher provide the mean and standard deviation for each of these three demographic characteristics? Why or why not? If the mean and standard deviation is not appropriate, what should the researchers report for that demographic characteristic?

SPSS QUESTION

1. Input the following set of scores into SPSS: 68, 9, 17, 87, 19, 37, 66, 78, 33, 29, 33, 61, 3, 57, 50. Calculate the range, standard deviation, and variance of these scores using SPSS.

JUST FOR FUN/CHALLENGE YOURSELF

1. Calculate the mean deviation of the following set of scores: 23, 45, 57, 62, 88.

2. What are all the possible scenarios in which the variance of a set of values would equal the standard deviation?

ANSWER KEY

TRUE/FALSE QUESTIONS

1. False. In fact, it is much more likely that these two sets of scores have different measures of variability.

2. True. This is possible, but it would be very rare. This would be the case only if the standard deviation and variance were both equal to 0 or 1.

3. True. The equation for both the standard deviation as well as the variance includes the mean, so this value must be calculated before the standard deviation or variance is calculated.

4. True. Both the mean and the standard deviation are sensitive to extreme scores, also known as outliers.

5. False. It is extremely rare for a set of scores to have zero variability.

6. False. This is because the standard deviation is stated in the original units from which it was calculated, while the variance is stated in squared units.

7. False. The standard deviation is the square root of the variance. If the variance is 6.78, then the square root of 6.78 is 2.60.

8. False. Given the extreme outlier ($60,000), the mean would be artificially inflated. The researcher would either want to discard that outlier or use the median (the central number in the data set) rather than the mean.

9. False. Ethnicity is a categorical, nominal variable. As such, the researcher would want to report the frequencies for each ethnicity (how many Caucasians, African Americans, Hispanics, etc.). Since you cannot find a mean for ethnicity, the mean is not appropriate. The standard deviation, which is derived from the mean, would similarly be inappropriate.

10. True.

11. False. The standard deviation is reported most frequently, and is the square root of the variance. Providing both is thus repetitive (as the variance can be easily calculated given the standard deviation, and vice versa).

MULTIPLE-CHOICE QUESTIONS

1. (d) 1, 4, 7, 9, 11

2. (c) 0, 2, 4, 6, 13

3. (d) 8, 8, 8, 8, 8

4. (a) The range

5. (b) Unbiased estimates

6. (c) The variance is equal to the square of the standard deviation.

7. (b) The standard deviation is the most widely used measure of variability in psychology.

8. (c) The mean represents the central score in each condition while the standard deviation represents the measure of spread in each condition.

9. (d) How much she donates only. This is a ratio scale, and thus has a mean. It is inappropriate to use the mean for the nominal rating of whether she donates or not.

EXERCISES

1. The range = 420 – 214 = 206.

2. First, $\bar{X} = \dfrac{\sum X}{n} \Rightarrow \dfrac{32+48+55+62+71}{5} = 53.6$.

Next, the standard deviation =

$$s = \sqrt{\frac{\sum(X-\bar{X})^2}{n-1}}$$

$$= \sqrt{\frac{(32-53.6)^2 +(48-53.6)^2 +(55-53.6)^2 +(62+53.6)^2 +(71-53.6)^2}{5-1}}$$

$$= 14.8.$$

Finally, the variance = $s^2 = 14.8^2 = 219.04$.

SHORT-ANSWER/ESSAY QUESTIONS

1. Unbiased estimates are preferred in situations in which a sample is used to estimate population parameters. In these situations, unbiased estimates give you a more conservative estimate than do biased estimates. For example, when 1 is subtracted from the sample size, the standard deviation is forced to be larger than it would be otherwise, giving a more conservative estimate.

2. Since the variance does not use the original scale as the original data, the variance score is rarely reported. In this case, the variance score exceeds the range of possible scores (1 to 7).

The researcher should report the standard deviation instead, taking the square root of the variance (in this case, the standard deviation is 3.3985). A standard deviation of 3.40 better represents the 1 to 7 fear-based scale.

3. The mean provides the measure of center, or the central number in a data set. The standard deviation provides the measure of spread, or the variability of the numbers in the data set.

4. The mean and standard deviation should be reported for the age variable only. Age is a continuous, ratio-oriented variable (ranging from 0 years old to over 100), and thus researchers can calculate the mean as well as the standard deviation. Ethnicity and gender, however, are nominal, categorical variables. Consider gender: there is no "average" gender (which would fall somewhere *between* males and females!). Frequencies are better reported for these two factors (the number of men and women; the number of Caucasians, African Americans, Hispanics, etc.).

SPSS QUESTION

Descriptive Statistics				
	N	Range	Std. Deviation	Variance
Var	15	84.00	27.51069	756.838
Valid *N* (listwise)	15			

JUST FOR FUN/CHALLENGE YOURSELF

1. Calculate the mean deviation of the following set of scores: 23, 45, 57, 62, 88.

First,

$$\bar{X} = \frac{\sum X}{n} \Rightarrow \frac{23+45+57+62+88}{5} = 55.$$

Next, the mean deviation

$$= \frac{\sum |X - \bar{X}|}{n}$$

$$= \frac{|23-55| + |45+55| + |57+55| + |62+55| + |88-55|}{5}$$

$$= 16.8.$$

2. For this to be the case, you would need the square of the standard deviation (i.e., the variance) to be equal to the standard deviation. In other words, you would need a number that does not change when it is squared. This leaves you with only two values: 0 and 1. The only two possible scenarios in which the variance would be equal to the standard deviation is when both of these values are 0 or when both of these values are 1.

4 A Picture Really Is Worth a Thousand Words

LEARNING OBJECTIVES

- Learn how to create visually appealing and useful representations of data.

- Review different ways in which data can be represented using tables and graphs.

- Be able to choose the best type of chart based on the nature of your data.

- Learn how to draw several graphs by hand, as well as how to create graphs using SPSS.

SUMMARY/KEY POINTS

- The two previous chapters focused on measures of central tendency and variability. This chapter expands upon this, illustrating how differences in these measures result in different-looking distributions.

- A visual representation of data can be much more effective than numerical values alone at illustrating the characteristics of a distribution or data set.
  - Tables can be used to illustrate the distribution of a variable. This can focus on a single factor (e.g., scores along an introversion/extroversion scale) or several factors (e.g., introversion/extroversion scores for both males and females).
  - Tables covered here include frequency distributions and cumulative frequency distributions. For example, a psychologist could look at the frequency of "supertasters" (people who are really sensitive to taste), "normal" tasters (most of the population), and "nontasters" (people who are relatively insensitive to taste) and plot the number of people in each category in a frequency distribution.
  - Charts/graphs can be used to pictorially represent the distribution of a variable. The vertical axis is the y-axis, and the horizontal axis is the x-axis. A good way to remember the difference is that the letter X stands on two legs, so it must "stand" on the horizontal axis. The letter Y, on the other hand, stands on one leg atop the vertical axis!
  - Charts/graphs covered here include:
    - histograms, which present numerically ordered quantitative categories (e.g., age or height might be in increasing or decreasing order along the x-axis). Here, the order is necessary, so the bars touch one another.
    - frequency polygons, which are virtually identical to the histogram except they include a continuous line representing the frequencies of scores within a class interval. (For example, think about height along the x-axis. There are few really short people and few really tall people, with most many people falling in-between. Connect the points at the top of each histogram with a line. For height, a bell-shaped curve often emerges!)
    - bar charts, which may look like a histogram at first, but use qualitative rather than quantitative data sets. (For example, think about dogs used in pet therapy. A researcher can see how many clients prefer poodles, Chihuahuas, corgis, French bulldogs, dachshunds, etc., along the x-axis. The researcher can place the various dog breeds in any order along the x-axis, as one breed is not quantitatively better than another). Here, the order of each category is unimportant, so bars do not touch one another.
    - column charts, which are similar to bar charts except the categories are placed along the y-axis and extend left to right (rather than up and down along the x-axis, as in bar charts)
    - line charts, which can show trends in data with equal intervals, like days of the week (e.g., the number of times an autistic child hits himself on the head on Monday, Tuesday, Wednesday, Thursday, Friday, Saturday, and Sunday).
    - pie charts, which focus on proportions (e.g., supertasters, normal tasters, and nontasters comprise 100% of the population, but normal tasters have a bigger slice of the pie!).

- It's easy to build a "bad" chart. Certain guidelines should be followed to make a chart that is easy to read and clearly illustrates what you are trying to show.
  - Less is more—minimize "chart junk." Avoid using fancy 3D charts, too many grid lines, etc.
  - Label everything.
  - Communicate only one idea.
  - Maintain the scale in a graph (in a 3:4 ratio).
  - A chart alone should convey what you want to say.
  - Simple is best; limit the number of words. Use the KISS principle: Keep It Simple, Student!

KEY TERMS

- **Frequency distribution:** A method of tallying and representing how often certain scores occur. Frequency distributions generally group scores into class intervals or ranges of numbers.

- **Class interval:** A range of numbers, chosen by the researcher, to be used in charts/graphs

- **Histogram:** A graphical representation of a frequency distribution in which the frequencies are represented by bars

- **Midpoint:** The central point of a class interval

- **Frequency polygon:** A continuous line that represents a frequency distribution

- **Cumulative frequency distribution:** A frequency distribution that shows frequencies for class intervals along with the cumulative frequency for each
  - **Ogive:** Another name for a cumulative frequency polygon

- **Column charts:** A type of chart in which categories are organized horizontally on the x-axis and values are shown vertically on the y-axis. This type of chart is used to compare the frequencies of different categories with one another.

- **Bar charts:** A type of chart in which categories are organized vertically on the y-axis and values are shown horizontally on the x-axis in the form of separate, separated bars

- **Line chart:** A type of chart in which categories are organized vertically on the y-axis and values are shown horizontally on the x-axis. These values are connected by one or more lines.

- **Pie chart:** A type of chart that illustrates the proportions of responses to an item as a series of wedges in a circle

TRUE/FALSE QUESTIONS

1. A bar chart tallies and represents how often certain scores occur in the form of a table.

2. Skewness is a measure of the central point of a class interval.

3. A frequency polygon can be defined as a continuous line that represents a frequency distribution.

4. When creating a graph, everything should be labeled, and the graph should only communicate one idea.

5. Graphs should contain as much text as possible.

6. Bar charts and histograms are the same thing.

7. A researcher studying the satisfaction that comes with success on a scale of 1 (not at all satisfied) to 9 (very satisfied) finds that participants with an intrinsic motivation (engaging in a task for enjoyment rather than a tangible reward) have higher levels of satisfaction than those with an extrinsic motivation (engaging in a task for a tangible reward, like money or a good grade). The best way to chart the satisfaction data is with a pie chart.

8. In a Milgram obedience study replication, a researcher has either a single participant or a group of participants provide shocks to someone who answered questions incorrectly. He then plots the amount of shock each gives along the x-axis. Here, the researcher can use two different lines in the same chart, one for the group participants and one for the single participants.

MULTIPLE-CHOICE QUESTIONS

1. When creating a graph, the ratio of the width to the length should be approximately
   _____.

   a. 1:2

   b. 2:1

   c. 3:4

   d. 5:7

2. When determining class intervals, you should aim to have about this many intervals cover the entire range of your data.

   a. 1 or 2

   b. 5 to 10

   c. 10 to 20

   d. 50 to 100

3. The largest class interval is placed _____ in a frequency distribution.

   a. at the top

   b. in the middle

   c. at the bottom

   d. randomly

4. When you want to compare the frequencies of different categories with one another, you should use a _____.

   a. pie chart

   b. line chart

   c. column chart

   d. frequency distribution

5. When you want to show a trend in the data at equal intervals, you should use a _____.

    a. pie chart

    b. line chart

    c. column chart

    d. frequency distribution

6. When you want to show the proportion of an item that makes up a series of data points, you should use a _____.

    a. pie chart

    b. line chart

    c. column chart

    d. frequency distribution

7. In a study looking at parent-child attachment relationships, a psychologist measures who helps more often when a child is crying: the child's mother or father. When reporting the data, the psychologist should use a _____.

    a. histogram

    b. line chart

    c. frequency polygon

    d. bar graph

8. A psychologist asks students to critique an opinion article in a campus newspaper. Half of the participants are told a faculty member wrote the article while others are told a student wrote the article. All participants then rate the extent to which they liked the article. Which kind of graph should the psychologist use to depict his results?

    a. Histogram

    b. Polygon

    c. Line chart

    d. Pie chart

9. To graph a qualitative variable (e.g., whether participants live in homes they own, homes they rent, hotels, or are homeless), use a _____ graph. To graph a quantitative variable (e.g., how much they like their home), use a _____ graph.

    a. pie; bar

    b. bar; histogram

    c. histogram; bar

    d. line; bar

EXERCISES

1. Draw a histogram using the data from the following frequency distribution:

Class Interval	Frequency
90–100	2
80–89	8
70–79	4
60–69	12
50–59	14
40–49	20
30–39	14
20–29	7
10–19	3
0–9	1

2. Make the histogram you just created into a frequency polygon.

3. Using the frequency distribution from question 1 in this section, add an additional column to make it into a cumulative frequency distribution.

SHORT-ANSWER/ESSAY QUESTIONS

1. Measures of central tendency and variability describe a group of data and how different scores are from each other. However, visual representations of data offer a more effective way to examine the characteristics of a distribution or data set. What are some different ways in which charts, graphs, and figures can illustrate data?

2. Different types of charts or graphs work better for different types of variables. What are three guidelines to keep in mind, or questions to ask yourself, when choosing which chart or graph to use?

3. You are studying the change over time in the number of course credits needed to graduate with a Bachelor of Arts degree. You have collected data on 3 universities at 10-year intervals over a 40-year span. What is the best chart or graph to illustrate your findings? Why?

4. You are interested in looking at the negative effects that high levels of lead in a local water supply have on the physical health of local community members. Using hospital records, you look at infants born within the last year, and you tally the number of boys and girls born as well as their weight at birth. You can then compare these rates to infants born a decade ago. What is the best chart to use for presenting infant gender, and what is the best chart to use for presenting infant birth weight? Defend your choices.

SPSS QUESTIONS

1. Create a histogram using the data from question 1 in the "Exercises" section.

2. Create a pie chart using the following data: Roman Catholic, 12 respondents; Protestant, 20 respondents; Jewish, 2 respondents; no religion, 6 respondents; other religion, 9 respondents.

JUST FOR FUN/CHALLENGE YOURSELF

1. Using the cumulative frequency distribution you came up with for question 3 under the "Exercises" section, chart the cumulative frequency data as a cumulative frequency polygon, or ogive.

ANSWER KEY

TRUE/FALSE QUESTIONS

1. False. This describes a frequency distribution.

2. False. This describes a midpoint.

3. True.

4. True. This is good advice to follow when creating a graph.

5. False. Including too many words can detract from the visual message your chart is intended to convey to readers.

6. False. The bars touch in a histogram, as the x-axis relies on interval data. The bars do not touch in the bar chart, as the x-axis relies on nominal or categorical data.

7. False. Satisfaction is a continuous variable along an interval scale. A line chart is more appropriate to show the trends in satisfaction, though a frequency polygon may also be used.

8. True. Researchers can have multiple lines in the same line chart.

MULTIPLE-CHOICE QUESTIONS

1. (c) 3:4

2. (c) 10 to 20

3. (a) at the top

4. (c) column chart

5. (b) line chart

6. (a) pie chart

7. (d) bar chart

8. (c) line chart

9. (b) bar; histogram

EXERCISES

1. Your histogram should look like this:

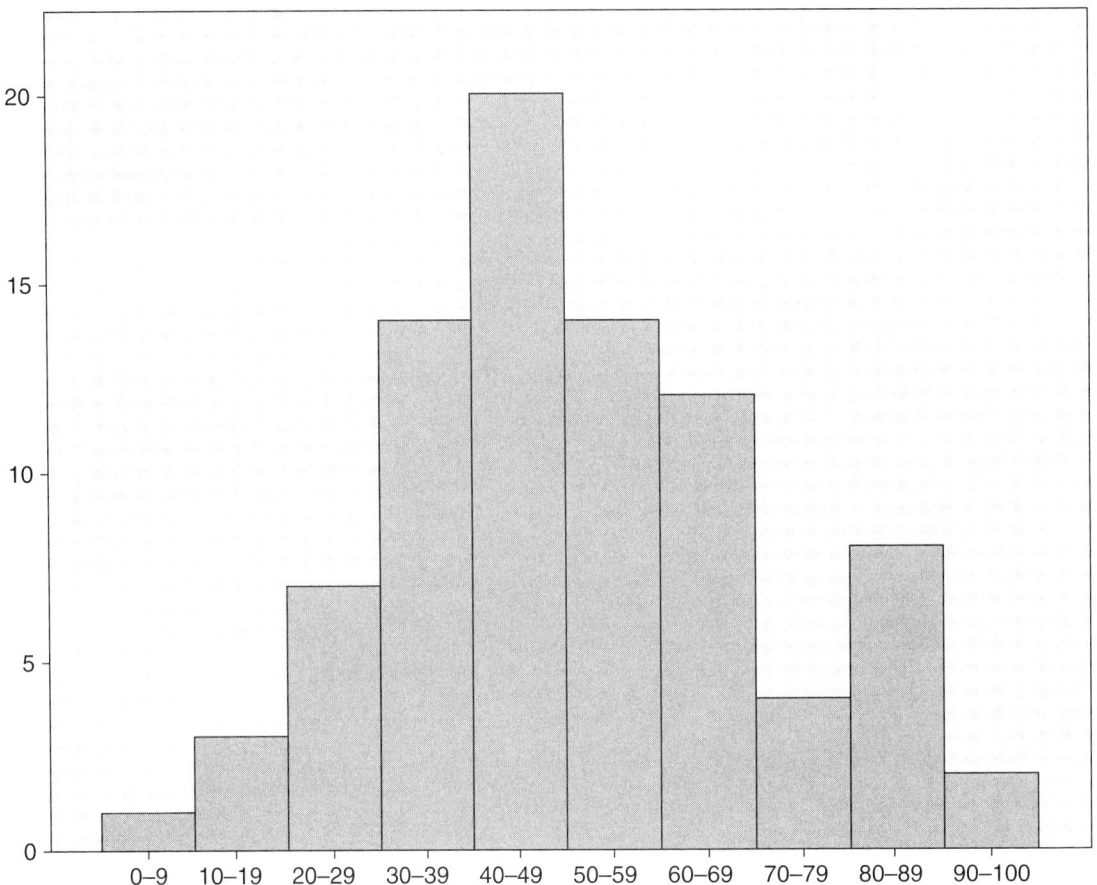

2. The frequency polygon should look like this:

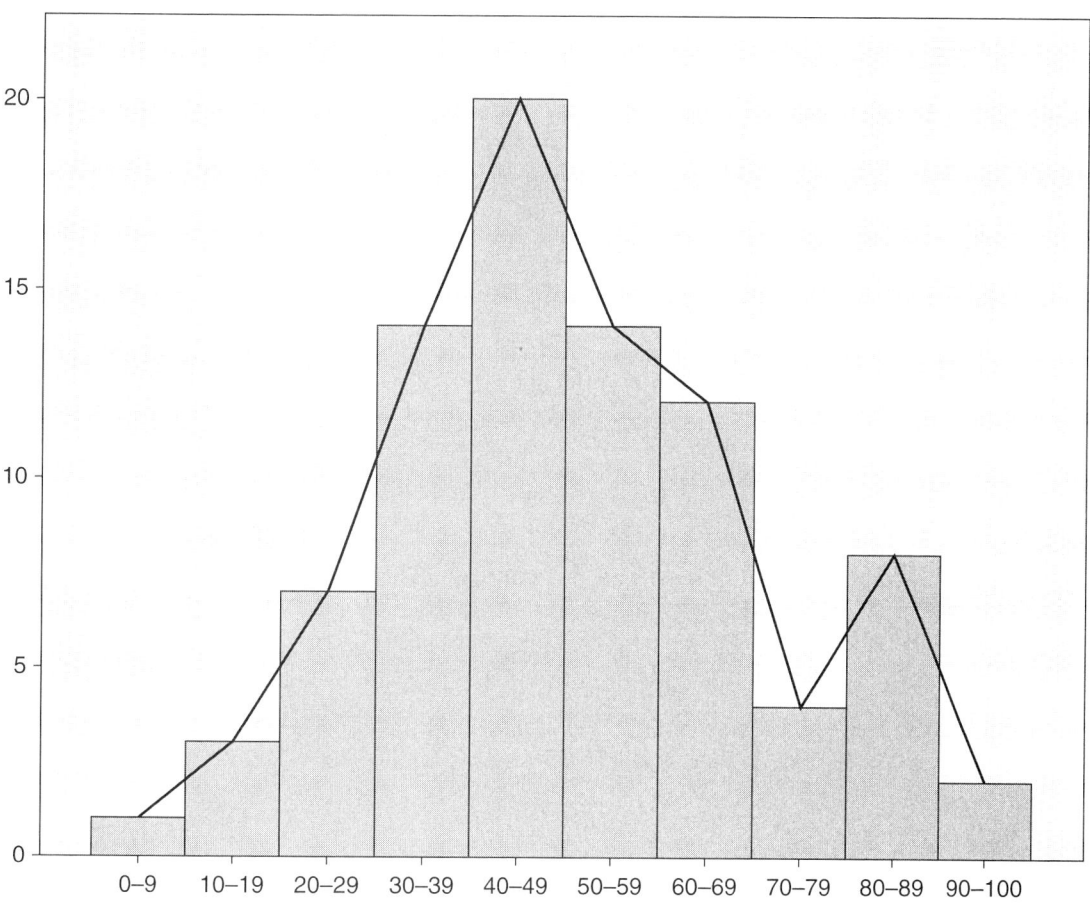

3. Your cumulative frequency distribution should look like this:

Class Interval	Frequency	Cumulative Frequency
90–100	2	85
80–89	8	83
70–79	4	75
60–69	12	71
50–59	14	59
40–49	20	45
30–39	14	25
20–29	7	11
10–19	3	4
0–9	1	1

SHORT-ANSWER/ESSAY QUESTIONS

1. Charts and graphs can illustrate the following:
   - That means and/or standard deviations have different distributions
   - What values of a variable occur and with what frequency
   - The data in a more dynamic manner than numbers alone can show
   - How much overlap exists between multiple distributions

2. Some of the questions that researchers may ask when determining how to illustrate their data include the following:
   - Is the measure of central tendency a mean, median, or mode?
   - Am I capturing one moment in time or telling a story about a trend for the same subjects over time?
   - Am I interested in showing the proportions of one category relative to others?
   - How many categories do I want to show at once?
   - Am I comparing the frequencies of different categories with one another?

3. A line chart is the best figure to illustrate your findings, because you can indicate the different schools with different lines and show changes in the number of credits (*y*-axis) at 10-year intervals (*x*-axis).

4. Gender is a categorical (or nominal) variable, and as such should be presented using either a pie chart or a bar chart. That way you can compare the proportion of male to female infants (in a pie chart) or show discrete categories (in a bar chart). Given the interval nature of birth weight, which can range from just a few pounds to many pounds, a histogram or line chart should be used.

SPSS QUESTIONS

1. See the "Exercises" section, question 1, for the correct histogram.

2. The pie chart should look like this:

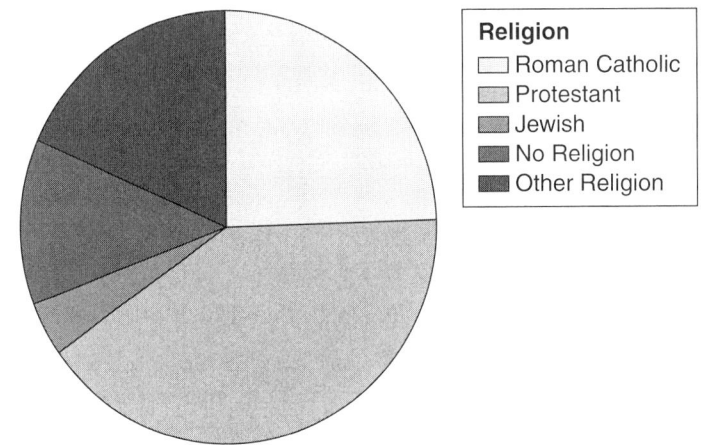

JUST FOR FUN/CHALLENGE YOURSELF

1. Your cumulative frequency polygon, or ogive, should look like the following image. Drawing the bars can be helpful if you are creating the cumulative frequency polygon by hand, but the bars are not necessary in an ogive.

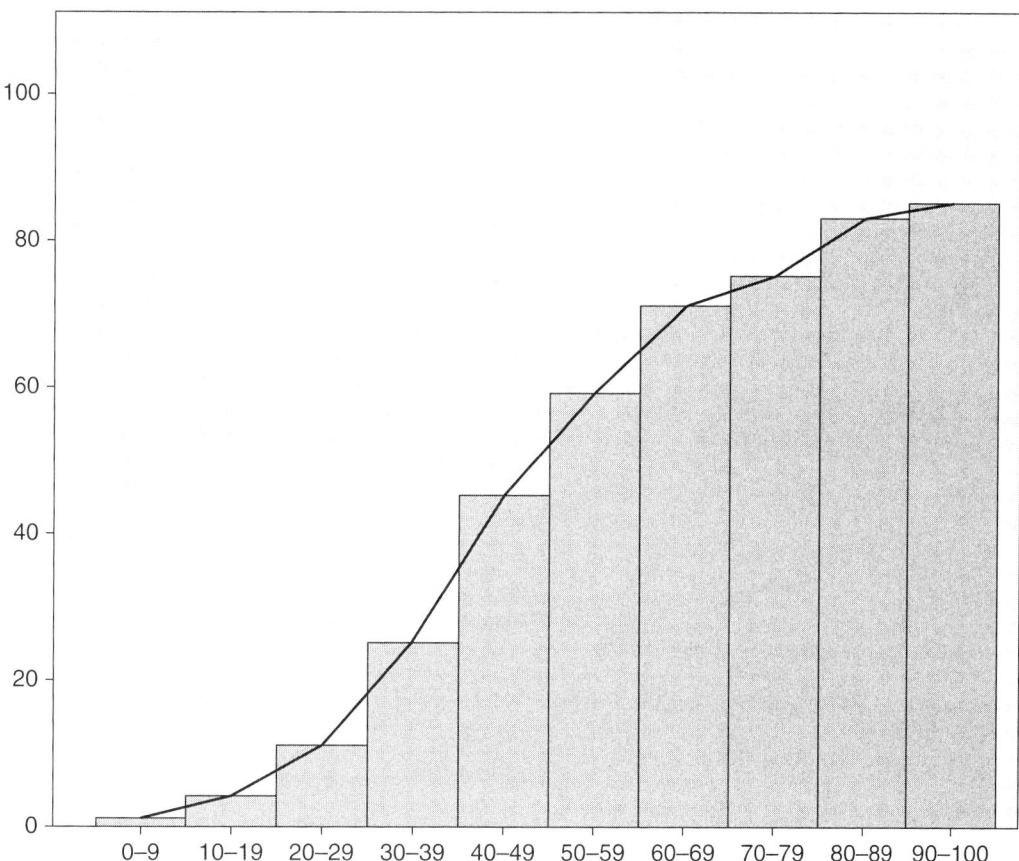

5 Ice Cream and Crime

Computing Correlation Coefficients

LEARNING OBJECTIVES

- Understand what correlation coefficients are used for.

- Learn how to interpret correlation coefficients.

- Be able to calculate Pearson's correlation coefficient by hand, as well as with SPSS.

- Be able to select the appropriate correlation coefficient to use depending on the nature of your variables.

SUMMARY/KEY POINTS

- Correlation coefficients are used to measure the strength and nature of the relationship between two variables.

- Pearson's correlation coefficient (r), the focus of this chapter, is used to calculate the correlation between two continuous (interval) variables.

- Other correlation coefficients can be used when one or more variables are ordinal or nominal.

- When you have a direct correlation—also known as a positive correlation—both variables change in the same direction. For example, as temperature increases, people feel more irritable. Conversely, as temperature decreases, people become less irritable. With an indirect correlation—also known as a negative correlation—variables change in opposite directions. For example, the more vaccinations a patient may get for a potential illness, the less likely they will get that illness.

- Correlation coefficients focus on generalities. This means that the correlation that you find describes the group, not every individual person in your data. Although there is a direct correlation between income and number of years spent in college, there are many people with PhDs who don't make a ton of money, and lots of high school dropouts who are multi-millionaires!

- The absolute value of the correlation coefficient reflects the strength of the correlation. Coefficients can range from -1 to $+1$, and the closer a coefficient's absolute value is to 1, the stronger the relationship. A correlation of .70 is not as strong as a correlation of $-.80$.

- The correlation between two variables will be reduced if the range of one or both of the variables is restricted.

- A scatterplot, or scattergram, can be used to visually illustrate a correlation between two variables. A positive slope represents a direct correlation, while a negative slope represents an indirect correlation.

- Correlation matrices are used to summarize correlations between a set of variables.

- The absolute value of a correlation corresponds to its strength—and corresponding meaningfulness—roughly as follows:
  - .8 to 1.0 is a very strong relationship.
  - .6 to .8 is a strong relationship.
  - .4 to .6 is a moderate relationship.
  - .2 to .4 is a weak relationship.
  - .0 to .2 is a weak or no relationship.

- The coefficient of determination is more precise than a correlation coefficient alone. The coefficient of determination is equal to the percentage of variance in one variable that is accounted for by the variance in a second variable.

- The fact that two variables are correlated does not imply that one causes the other. For example, there is a direct, positive correlation between hair length and success as a psychology student. This doesn't mean that long hair causes such success. There might be other factors that underlie the relationship. Gender, after all, might factor into this relationship, with females having both longer hair and higher psychology grades than males.

KEY TERMS

- **Correlation coefficient**: A numerical index that reflects the relationship between two variables
  - Its range is from -1 to $+1$.
  - It is also known as a bivariate correlation.

- **Pearson product-moment correlation**: A specific type of correlation coefficient developed by Karl Pearson. It is specifically suited to determining the correlation between two continuous variables.

- **Direct correlation**: A positive correlation such that the values of both variables change in the same direction

- **Indirect correlation**: A negative correlation such that the values of the two variables move in opposite directions

- **Scatterplot** or **scattergram**: A plot of matched data points. This type of chart is used to illustrate a correlation between two variables.

- **Linear correlation**: A correlation that is best expressed as a straight line

- **Curvilinear relationship**: A situation in which the correlation between two variables begins as a direct correlation and then becomes an indirect correlation, or vice versa. It can be detected by examining the scatterplot.

- **Correlation matrix**: A table of correlation coefficients in which variables comprise the rows and columns of the table and the intersections of the variables are represented by correlation coefficients

- **Coefficient of determination**: The amount of variance accounted for in a relationship between two variables
  - It is equal to the square of the Pearson product-moment correlation coefficient.

- **Coefficient of alienation** (aka coefficient of nondetermination): The amount of unexplained variance in a relationship between two variables
  - It is equal to 1 minus the coefficient of determination.

- **Phi coefficient**: A measure used to estimate the correlation between two nominal variables

- **Rank biserial coefficient**: A measure used to estimate the correlation between one nominal and one ordinal variable

- **Point biserial coefficient**: A measure used to estimate the correlation between one nominal and one interval variable

- **Spearman rank coefficient**: A measure used to estimate the correlation between two ordinal variables

TRUE/FALSE QUESTIONS

1. If two variables are correlated, then one of the variables causes the other.

2. If your variables are found to be correlated, then the variables are correlated for the entire group of respondents but most likely not correlated for each individual case in your data set.

3. Arriving at a negative correlation is always a worse result than finding a positive correlation.

4. A correlation of −.33 is stronger than a correlation of .21.

5. A psychologist finds that the more time that withdrawn children spend with therapy dogs, the more social the children become. This represents a positive correlation.

MULTIPLE-CHOICE QUESTIONS

1. Which of the following is the range for Pearson's correlation coefficient?

   a. −10 to +10

   b. −5 to +5

   c. −1 to +1

   d. −100 to +100

2. Pearson's product-moment correlation can be used to calculate the correlation between these two types of variables.

   a. Two interval variables

   b. Two ordinal variables

   c. One nominal and one ordinal variable

   d. Two nominal variables

   e. All of the above

3. If one variable increases while the other decreases, you have this type of correlation.

   a. Direct correlation

   b. Indirect correlation

   c. Curvilinear correlation

   d. Leptokurtic correlation

4. If two variables move in the same direction (i.e., one increases as the other increases, and one decreases as the other decreases), you have a _____ correlation.

   a. direct

   b. indirect

   c. curvilinear

   d. leptokurtic

5. Which of the following is the strongest correlation?

   a. −.15

   b. +.27

   c. −.70

   d. +.55

6. Which of the following is the weakest correlation?

   a. −.22

   b. −.78

   c. +.12

   d. +.89

7. Pearson's product-moment correlation coefficient is represented by which of the following letters?

   a. *r*

   b. *p*

   c. *t*

   d. *c*

   e. *z*

8. If you compute the correlation between two variables, and one of the variables never changes, you can be sure that the Pearson correlation coefficient is equal to _____.

   a. +1

   b. 0

   c. −1

   d. +.5

   e. −.5

9. If a correlation is computed between two variables, but the range of one of the variables is restricted, your correlation will be _____.

   a. lower

   b. higher

   c. the same

   d. 0

   e. +1

10. In a scatterplot, if the dots cluster from the lower left-hand corner to the upper right-hand corner, the two variables have _____ correlation.

    a. a direct

    b. an indirect

    c. a curvilinear

    d. a bilinear

11. Which of the following would be considered a very strong correlation coefficient?

    a. .8 to 1.0

    b. .6 to .8

    c. .4 to .6

    d. .2 to .4

    e. .0 to .2

12. Which of the following would be considered a moderate correlation coefficient?

    a. .8 to 1.0

    b. .6 to .8

    c. .4 to .6

    d. .2 to .4

    e. .0 to .2

13. View the following scatterplot. What is the best estimate of the correlation between these two variables?

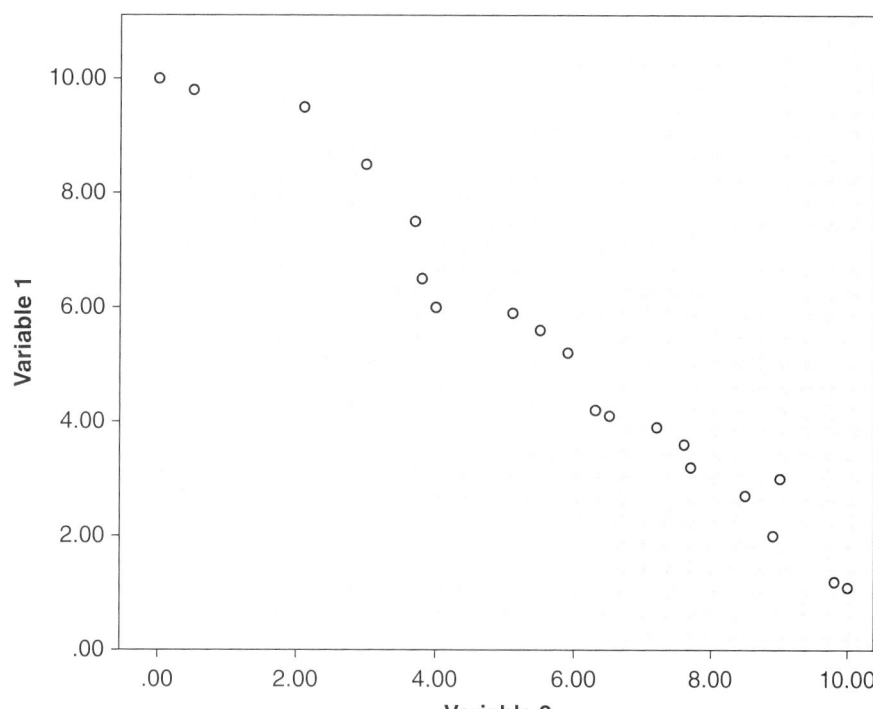

    a. −.20

    b. +.55

    c. −.98

    d. +.90

14. Which of the following correlation coefficients would indicate a weak or no relationship between two variables?

   a. .8 to 1.0

   b. .6 to .8

   c. .4 to .6

   d. .2 to .4

   e. .0 to .2

15. Which correlation would you use when analyzing the relationship between two nominal variables?

   a. Pearson's product-moment correlation coefficient

   b. Spearman rank coefficient

   c. Point biserial

   d. Rank biserial coefficient

   e. Phi coefficient

16. Which correlation would you use when analyzing the relationship between a nominal and an interval variable?

   a. Pearson's product-moment correlation coefficient

   b. Spearman rank coefficient

   c. Point biserial

   d. Rank biserial coefficient

   e. Phi coefficient

17. Which correlation would you use when analyzing the relationship between two ordinal variables?

   a. Pearson's product-moment correlation coefficient

   b. Spearman rank coefficient

   c. Point biserial

   d. Rank biserial coefficient

   e. Phi coefficient

18. A researcher obtains a correlation of $r = -.72$ between grade point average (GPA) and time spent watching television for a sample of college students. For this example, who tends to get the better grades?

   a. Students who watch a lot of television

   b. Students who do not watch a lot of television

   c. Students do not score significantly different regardless of how much TV they watch

   d. Both students who watch a lot of television and those who do not watch a lot of television have low grades

19. Research shows that the more the temperature goes up, the more irritable people become. This best represents what kind of correlation?

    a. Positive (direct) correlation

    b. Negative (indirect) correlation

    b. No correlation

    d. There is not enough information to decide if there is a correlation

20. Consider the graph below. There is _____ correlation between calories eaten daily and weight.

    a. a positive (direct)

    b. a weak

    c. a negative (indirect)

    d. no

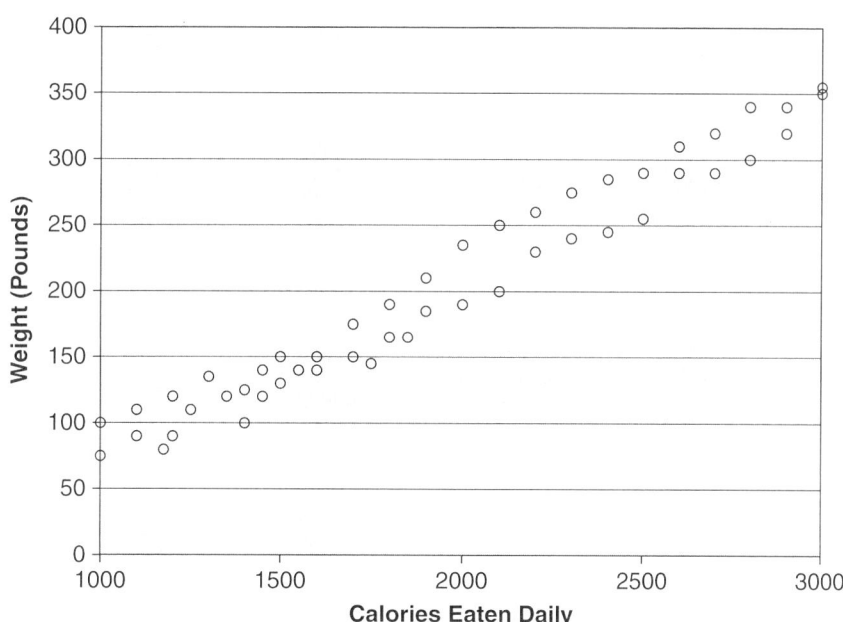

21. James conducts a study on the correlation between the number of churches and amount of crime in a county and finds a strong positive relationship between the two (in other words, the more churches there are in a county, the more crime there is). What should he conclude?

    a. Churches cause crime; therefore, cities should attempt to reduce the number of churches.

    b. Committing a crime causes people to go to church more often, probably in an attempt to atone for their sins.

    c. High population density causes both churches and crime to increase in any geographical location.

    d. None of the above; it is not appropriate to make causal claims based on correlational data.

EXERCISES

1. Compute the Pearson product-moment correlation coefficient for the following two variables:

Years of Education	Income (000)
8	12
9	15
11	14
12	20
14	28
14	33
16	33
16	45
18	48
22	61
23	81

2. You find that the correlation between two variables is equal to +.76. First, judge the sign (+ or −) and strength of the correlation. Is it direct or indirect? Now, calculate the coefficient of determination and the coefficient of alienation. What do both of these values mean?

3. Imagine you conduct a study looking at the correlation between teenagers' overall satisfaction in their relationship and how happy they feel about the relationship. Use the table below to compute the Pearson correlation.

Relationship Satisfaction	Relationship Happiness
4	3
6	7
3	3
4	5
7	6
3	2
5	7

SHORT-ANSWER/ESSAY QUESTIONS

1. To determine the relationship between two continuous variables, you run a correlation in SPSS, expecting to find a sizable positive correlation between the variables. However, the SPSS output shows the Pearson's *r* is small. Should you give up on your hypothesis about the relationship between these variables? How else could you examine the data? What might a figure tell you about the relationship of your variables?

2. Correlations do not represent causality. Therefore, how could you describe the value of using a correlation to examine your data?

3. Three main types of correlation coefficients are described in this chapter (i.e., correlation coefficient, coefficient of determination, and coefficient of alienation). Please describe the identifying features of each.

4. A study by Levesque (1993) looked at teenager relationship satisfaction ratings and correlated it with several features of that relationship. He broke down the correlations based on gender of the teenager. Given the table below, what conclusions can you draw for each feature of the relationship?

	Correlation of Satisfaction	
Features of a Relationship	**Males**	**Females**
Togetherness	.48*	.30*
Personal Growth	.44*	.22*
Appreciation	.33*	.21
Exhilaration/Happiness	.46*	.39*
Painfulness/Emotional Turmoil	19	−.09
Passion/Romance	−.09	.21*
Emotional Support	.34*	.23*
Good Communication	.13	.17

* Indicates significant at the p < .05 level

5. A researcher finds that increasing the number of tasks a participants must complete in one hour increases the frustration of the participant. If the correlation between number of tasks and frustration levels is .70, what percent of the variance is shared (i.e. the coefficient of determination)? What other factors might account for the remaining nonshared variance? List a few possible factors.

SPSS QUESTIONS

1. Input the data given for question 1 of the "Exercises" section into SPSS. Now, calculate the correlation coefficient using SPSS. Does your result match what you calculated by hand?

2. Create a scatterplot in SPSS using the same data. What does it show?

JUST FOR FUN/CHALLENGE YOURSELF

1. Describe a curvilinear relationship. Come up with an example of two variables that may have a curvilinear relationship and explain why the relationship is curvilinear (use an example other than the one that was presented in the book).

2. If you are computing correlation coefficients among 10 variables, how many unique correlation coefficients will you calculate in total?

ANSWER KEY

TRUE/FALSE QUESTIONS

1. False. Even if two variables are correlated with each other, one variable does not necessarily cause the other. The example presented in this chapter's discussion of the correlation between ice cream consumption and crime illustrates this possibility very well.

2. True. Correlations are conducted on your entire set of data: This means that, while any correlations that you find are true for the entire data set, they're not necessarily true for each individual case or person. For example, if you found a direct (positive) correlation between years of education and income, this would not necessarily mean that all individuals with low education have low income (think Bill Gates). In practice, you will typically find some cases that diverge from the correlation.

3. False. A negative correlation, on its own, is neither any better nor any worse than a positive correlation.

4. True. Because the strength of the correlation is based on how close a number is to -1 or $+1$, a correlation of $-.33$ is closer to -1 than a correlation of $.22$ is to $+1$.

5. True. As one variable increases, the other also increases, resulting in a positive correlation.

MULTIPLE-CHOICE QUESTIONS

1. (c) -1 to $+1$
2. (a) Two interval variables
3. (b) Indirect correlation
4. (a) direct
5. (c) $-.70$
6. (c) $+.12$
7. (a) r
8. (b) 0
9. (a) lower
10. (a) a direct
11. (a) .8 to 1.0
12. (c) .4 to .6

13. (c) −.98

14. (e) .0 to .2

15. (e) Phi coefficient

16. (c) Point biserial

17. (b) Spearman rank coefficient

18. (b) Students who do not watch a lot of television

19. (a) Positive correlation

20. (a) a positive

21. (d) None of the above; it is not appropriate to make causal claims based on correlational data.

EXERCISES

1.

	Years of Ed	Income	X²	Y²	XY
	8	12	64	144	96
	9	15	81	225	135
	11	14	121	196	154
	12	20	144	400	240
	14	28	196	784	392
	14	33	196	1,089	462
	16	33	256	1,089	528
	16	45	256	2,025	720
	18	48	324	2,304	864
	22	61	484	3,721	1,342
	23	81	529	6,561	1,863
Totals:	163	390	2,651	18,538	6,796

$$r_{XY} = \frac{n\Sigma XY - \Sigma X \Sigma Y}{\sqrt{(n\Sigma X^2 - (\Sigma X)^2)(n\Sigma Y^2 - (\Sigma Y)^2)}}$$

$$= \frac{11 \times 6{,}796 - 163 \times 390}{\sqrt{(11 \times 2{,}651 - 163^2)(11 \times 18{,}538 - 390^2)}}$$

$$= \frac{74{,}756 - 63{,}570}{\sqrt{(2{,}592)(51{,}818)}}$$

$$= .9652$$

2. First, this correlation coefficient was found to be positive, meaning that this is a direct correla-tion. This means that the two variables included in the analysis tend to "move together"; in other words, as one variable increases, the other is expected to increase, and as one variable decreases, the other is expected to decrease. Next, as the correlation was found to be +.76, you could state that there is a strong relationship between these two variables.

Next, the coefficient of determination is simply calculated as the square of the correlation coefficient. In this example, the coefficient of determination is equal to .58. The coefficient of alienation is equal to 1 minus the coefficient of determination, or $1 - .58 = .42$. From the coefficient of determination, it can be said that 58% of the variance in one of the variables is explained by the variation in the second variable. The coefficient of alienation means that 42% of the variance in either of the variables cannot be explained by the other variable.

3.

	Satisfaction	Happiness	X²	Y²	XY
	4	3	16	9	12
	6	7	36	49	42
	3	3	9	9	9
	4	5	16	25	20
	7	6	49	36	42
	3	2	9	4	6
	5	7	25	49	35
Totals:	32	33	160	181	166

$$r_{xy} = \frac{n\Sigma XY - \Sigma X \Sigma Y}{\sqrt{[n\Sigma X^2 - (\Sigma X)^2][n\Sigma Y^2 - (\Sigma Y)^2]}}$$

$$r_{xy} = \frac{7(166) - (32 \times 33)}{\sqrt{[7(160) - 32^2][7(181) - 33^2]}}$$

$$r_{xy} = \frac{1162 - 1056}{\sqrt{[1120 - 1024][1267 - 1089]}}$$

$$r_{xy} = \frac{106}{\sqrt{[96][178]}}$$

$$r_{xy} = \frac{106}{130.72}$$

$$r_{xy} = \frac{106}{\sqrt{17088}}$$

$$r_{xy} = .8108$$

SHORT-ANSWER/ESSAY QUESTIONS

1. No, the Pearson's r correlation does not tell the whole story. The next thing to look at is a scat-terplot/scattergram. If the dots on the scatterplot are in no particular order, then you can accept the value of the correlation. If, however, the dots on the scatterplot show either a U shape or an inverted (upside-down) U shape, that would indicate a curvilinear relationship. This means that as one continuous variable changes, the relationship with another continuous variable either gets stronger, plateaus, and then gets weaker or the opposite—the relationship with the other continuous variable gets weaker, stays stable, and then gets stronger again.

2. Even if neither variable can be said to cause a change in the other, the correlation still conveys important messages. A correlation indicates the strength of a relationship between two contin-uous variables. It can further show a positive relationship (the variables change in the same direction) or a negative relationship (as one variable goes up, the other variable goes down). The coefficient of determination can also show the amount of variability in one variable that

is accounted for by the variability in the other variable. All of this information can give the researcher ideas about how much of the change in a variable potentially remains to be described by other variables.

3. A correlation coefficient is a number—numerical index—that reflects the relationship between two variables. The coefficient of determination is the percentage of the variance in one variable that is accounted for by the variance in the other variable. The coefficient of alienation/nonde-termination is the amount of variance in one variable left over and *not* accounted for by the variance in the other variable.

4. Both males and females who rate togetherness, personal growth, exhilaration/happiness, and emotional support have higher satisfaction in relationships (all are strong, positive correlations). However, males who rate appreciation highly have high satisfaction in their relationships (this feature is not significant for females). Females who rate passion/romance high have significantly higher satisfaction in their relationships (this feature is not significant for males).

5. The coefficient of determination (shared variance) is 49%, or .70 × .70. There might be several factors that account for the nonshared variance (coefficient of nondetermination, or alienation), including: the individual's own temperament, the age of the participant (maybe older people are less able to cope with multiple tasks), familiarity with research studies, and whether the participant expects an easy or difficult research task, among many others.

SPSS QUESTIONS

1. The following table presents the correct SPSS output. As you can see, the correlation coefficient matches the one calculated by hand.

Correlations

		Education	**Income (1,000s)**
Education	Pearson Correlation	1	.965**
	Sig. (2-tailed)		.000
	N	11	11
Income (1000s)	Pearson Correlation	.965**	1
	Sig. (2-tailed)	.000	
	N	11	11

** Correlation is significant at the 0.01 level (2-tailed).

2. The following figure presents this scatterplot. Based on the scatterplot, there appears to be a direct (or positive) correlation, with a strong relationship between these two variables.

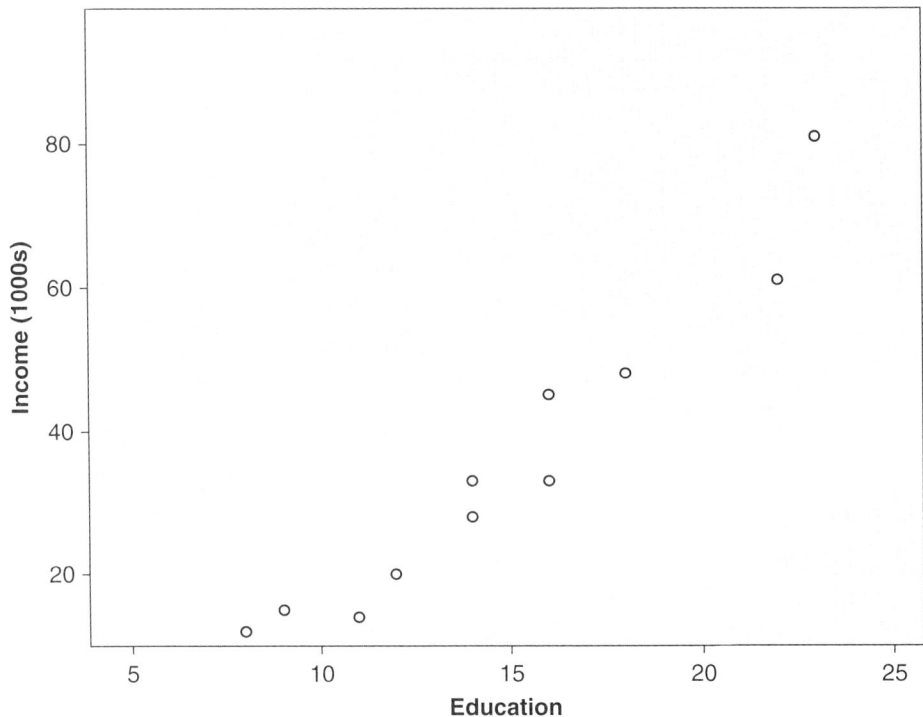

JUST FOR FUN/CHALLENGE YOURSELF

1. In a curvilinear relationship, the nature of the relationship between two variables substantially changes over the range of these variables. Specifically, a direct, or positive, correlation between two variables becomes an indirect, or negative, relationship (or vice versa). There are many possible examples. One example could be the relationship between work satisfaction and hours worked per week. To illustrate, individuals who only work a few hours per week may be very dissatisfied with their work, because their income is very low and they're not able to work as many hours as they would like. However, individuals who work very long hours— for example, 70 to 80 hours a week or more—may also be very dissatisfied with their work due to stress and a lack of free time. Individuals who work around a normal workweek, 40 hours per week, could have the greatest levels of job satisfaction, because they both make enough income and have enough free time. If this relationship were plotted on a graph, it would be in the shape of an upside-down U: Individuals with the fewest hours also have the lowest satisfaction. Then satisfaction increases as hours per week increase, up to around 40 hours per week. Beyond this point, satisfaction decreases as hours per week continue to increase. When hours per week are low, the relationship between these two variables is direct. However, when hours per week are high, the relationship between these two variables is indirect.

2. This can be calculated in the following way:

Number of unique correlations $\dfrac{n(n-1)}{2} \Rightarrow \dfrac{10(10-1)}{2} = 45.$

6 Just the Truth

An Introduction to Understanding Reliability and Validity

LEARNING OBJECTIVES

- Understand the difference between reliability and validity and learn why they are important.

- Learn the difference between the different types of reliability and validity.

- Understand what steps you can take if you need to increase reliability or validity.

- Learn how to compute and interpret different types of reliability and validity, both by hand and with SPSS.

SUMMARY/KEY POINTS

- Before we start analyzing and interpreting data, it will be important to understand what reliability and validity are, because they both have very important implications for the data.
  - Reliability explores this question: How do I know that the test, scale, instrument, etc., I use works every time I use it? In other words, can it be repeated?
  - Validity explores this question: How do I know that the test, scale, instrument, etc., I use measures what it is supposed to?
  - If data are not reliable or not valid, then the results of any test or hypothesis will have to be inconclusive, as you are unsure about the quality of the data.
  - Something can be reliable without being valid. For example, if I give participants ten circles and ten squares and ask them to pick out all of the circles, they could easily pick out all of the squares. Future participants might do the same thing. This would make it reliable (they always pick out the squares), but not valid (since I asked them to pick out circles).

- A dependent variable is the outcome or predicted variable in an analysis, while an independent variable is the treatment or predictor variable in an analysis. An independent variable might be manipulated. For example, if I wanted to manipulate participants' self-esteem levels, I might tell some that they failed a task while I tell others that they passed it, even if I never look at the actual results. An independent variable might also be measured. I can see if males or females differ in self-esteem, but I cannot manipulate the participant's gender.

- Reliability relates to the degree to which a test measures something consistently.
  - Observed scores are the actual or measured scores, while true scores are the scores research participants would receive if the test contained no error.
    - The difference between these two scores is called the error score.
    - Observed scores may come close to true scores, but in the social and behavioral sciences, they are rarely the same as true scores due to the presence of error. Imagine you give someone a warning versus no warning about where a quickly flashed image will appear on their computer screen. You then measure how quickly they respond. You might conclude that a warning helps them spot the image more quickly than no warning, but there are things other than the warning that might influence participant reflexes, each of which contribute error to the study (e.g., their familiarity with computers, how tired they might be, whether their mind is focused on another task, etc.).
    - The less error you have, the greater the reliability.
  - Types of reliability include test-retest reliability, parallel forms reliability, internal consistency reliability, and interrater reliability.
    - Test-retest reliability is calculated as the correlation between scores from time 1 and scores from time 2. For example, if you take the same test back-to-back, your first score should be comparable to your second score. If you do well at time 1, you should do well at time 2.
    - Parallel forms reliability is calculated as the correlation between scores from the first form of your test with the scores from the second form of your test. Here, the two forms of the test differ. They might parallel each other in terms of content, but the actual questions differ.
    - Internal consistency reliability is calculated using Cronbach's alpha.
    - Interrater reliability is calculated as the number of agreements between your two raters divided by the total number of possible agreements. If two researchers observing aggression on the playground both see aggression during 8 out of 10 observation sessions, their agreement is 80%.
    - For high reliability, you want your reliability coefficients to be positive and to be as large as possible (1 is the highest possible reliability coefficient score while .00 is the lowest possible score).
    - If your test is not reliable, you must try to lower your error. A number of possible changes can be made for this purpose, including increasing the number of items or observations and deleting unclear items.

- Validity is the degree to which an assessment tool measures what it says it does.
  - Measures of validity include content validity, criterion validity, and construct validity.
    - Content validity is established by consultation with a content expert on the topic your instrument focuses on.
    - Criterion validity is determined on the basis of the association between the test scores and some specified present or future criterion.
    - Construct validity is based on a judgment of how well your test reflects an underlying construct or idea.
  - If a test or instrument lacks validity, changes can be made to improve validity. What changes are appropriate to make will depend on the type of validity in question.

- ○ Content validity: Redo questions so that, according to an expert judge, they more accurately reflect the topics and ideas being tested for.
- ○ Criterion validity: Reexamine the nature of the items on the test and question whether responses should be expected to relate to the criterion you selected.
- ○ Construct validity: Review the theoretical rationale that underlies your test.
- ○ When working on a thesis or dissertation, it is strongly recommended that an instrument or text be used that has already been established to be reliable and valid.
- ○ A test can be reliable and not valid, but it is impossible to have a valid test that is not reliable.
  - ○ The maximum level of validity possible is equal to the square root of the reliability coefficient.

- For a psychological example, consider the Ambivalent Sexism Inventory (ASI) created by Fiske and Glicke. This 22-item scale asks 11 questions related to hostile sexism (which measures participants' antagonistic attitudes toward women, who are often viewed as trying to control men through feminist ideology or sexual seduction) and 11 questions related to benevolent sexism (which measures participants' chivalrous attitudes toward women; although this might seem favorable, it is sexist in the sense that it casts women as weak creatures in need of men's protection). But the ASI began with 140 statements, some good and some bad. Through several administrations of the survey, the authors weeded out the bad statements to see which yielded reliable (consistent) results across administrations, and they assessed the validity of the scales themselves (did the hostile and benevolent sexism questions *really* focus on hostile and benevolent sexism, respectively?). The final group of questions showed high internal consistency as well. For example, Chronbach's alpha for hostile sexism questions showed high correlations between questions, ranging from .80 to .92. The authors have validated the scale across many different populations throughout the world and at different times. As you can imagine, it was very time-consuming and laborious, but it is now a common scale used in studies looking at workplace interactions between males and females.

Glick, P., & Fiske, S. T. (1996). The ambivalent sexism inventory: Differentiating hostile and benevolent sexism. *Journal of Personality and Social Psychology, 70*(3), 491–512

KEY TERMS

- **Dependent variable**: The outcome of or predicted variable in an analysis

- **Independent variable**: A treatment variable that is manipulated or the predictor variable in an analysis

- **Reliability**: The degree to which a test measures something consistently
  - ○ **Observed score**: The actual score that is recorded or observed
  - ○ **True score**: The score that you would receive if a test measured your ability perfectly
  - ○ **Error score**: The part of a test score that is random and contributes to the unreliability of a test
  - ○ **Test-retest reliability**: A type of reliability that examines consistency over time
  - ○ **Parallel forms reliability**: A type of reliability that examines consistency across different forms of the same test
  - ○ **Internal consistency reliability**: A type of reliability that measures the extent to which items on a test are consistent with one another and represent one—and only one—dimension, construct, or area of interest
  - ○ **Cronbach's alpha**: A particular measure of internal consistency reliability
  - ○ **Interrater reliability**: A type of reliability that examines the consistency of raters

- **Validity**: The quality of a test such that it measures what it says it does
  - ○ **Content validity**: A type of validity that examines the extent to which a test accurately reflects all possible topics and ideas under the subject or topic you are testing for
  - ○ **Criterion validity**: A type of validity that examines how well a test reflects some criterion that occurs in the present or future
    - ○ **Concurrent criterion validity**: A type of validity that examines how well a test outcome is consistent with a criterion that occurs in the present
    - ○ **Predictive validity**: A type of validity that examines how well a test outcome is consistent with a criterion that occurs in the future
  - ○ **Construct validity**: A type of validity that examines how well a test reflects an underlying construct or idea

TRUE/FALSE QUESTIONS

1. Reliability and validity should be determined after an analysis is already complete.

2. If a test or instrument contains no error, then it can be said to have perfect reliability.

3. Observed scores are commonly exactly the same as true scores.

4. Smaller amounts of error are associated with greater reliability.

5. High reliability is associated with small reliability coefficients, as close as possible to –1.

6. If reliability is very low, nothing needs to be done before reporting this finding in your paper.

7. When working on your thesis or dissertation, it is best to create your own test instrument.

8. If a psychologist finds high internal consistency (a Cronbach's alpha of .811) among several questions looking at happiness, this means the happiness questions tend to measure the same construct.

9. Two psychologists observe the number of times a parent spanks a child in a shopping mall setting in a 12-hour time span. They count whether they observe spanking at least once during each hour. One observer counts 12 incidents (at least one in each hour) while the other only observes 7 incidents (in 7 of the 12 hours). The interater reliability between the two observers is .66.

10. It is possible for a study to have high validity but low reliability.

MULTIPLE-CHOICE QUESTIONS

1. The manipulated treatment, or predictor variable in an analysis, is known as _____ variable.

   a. the dependent

   b. the independent

   c. the correlated

   d. a skewed

2. The outcome, or predicted variable in an analysis, is known as _____ variable.

    a. the dependent

    b. the independent

    c. the correlated

    d. a skewed

3. Reliability serves to answer which of the following questions?

    a. How do I know that the test, scale, instrument, etc., I use measures what it is supposed to?

    b. How do I know that the test, scale, instrument, etc., I use works on all populations?

    c. How do I know that the test, scale, instrument, etc., I use works every time I use it?

4. Validity serves to answer which of the following questions?

    a. How do I know that the test, scale, instrument, etc., I use works every time I use it?

    b. How do I know that the test, scale, instrument, etc., I use works on all populations?

    c. How do I know that the test, scale, instrument, etc., I use measures what it is supposed to?

5. The actual, or measured, score is called the _____ score.

    a. true

    b. observed

    c. measured

    d. error

    e. perfect

6. The score that you would receive if a test contained no error is called the _____ score.

    a. true

    b. observed

    c. measured

    d. error

    e. perfect

7. The difference between the score received if a test contained no error and the actual/measured score is called the _____ score.

    a. true

    b. observed

    c. measured

    d. error

    e. perfect

8. Test-retest reliability examines which of the following?

    a. The consistency of raters

    b. The consistency across different forms of the same test

    c. The extent to which items in a test are consistent with one another and represent exactly one dimension, construct, or area of interest

    d. Consistency over time

9. Parallel forms reliability examines which of the following?

    a. The consistency of raters

    b. The consistency across different forms of the same test

    c. The extent to which items in a test are consistent with one another and represent exactly one dimension, construct, or area of interest

    d. Consistency over time

10. Internal consistency reliability examines which of the following?

    a. The consistency of raters

    b. The consistency across different forms of the same test

    c. The extent to which items in a test are consistent with one another and represent exactly one dimension, construct, or area of interest

    d. Consistency over time

11. Interrater reliability examines which of the following?

    a. The consistency of judges

    b. The consistency across different forms of the same test

    c. The extent to which items in a test are consistent with one another and represent exactly one dimension, construct, or area of interest

    d. Consistency over time

12. _____ reliability is calculated using Cronbach's alpha.

    a. Test-retest

    b. Parallel forms

    c. Internal consistency

    d. Interrater

13. _____ reliability is calculated as the number of agreements between judges divided by the total number of possible agreements.

    a. Test-retest

    b. Parallel forms

    c. Internal consistency

    d. Interrater

14. _____ reliability is calculated as the correlation between scores from time 1 and scores from time 2.

   a. Test-retest

   b. Parallel forms

   c. Internal consistency

   d. Interrater

15. _____ reliability is calculated as the correlation between scores from the first form of the test with the scores from the second form of the test.

   a. Test-retest

   b. Parallel forms

   c. Internal consistency

   d. Interrater

16. _____ validity examines how well a test reflects some standard that occurs in the present or future.

   a. Predictive

   b. Concurrent criterion

   c. Criterion

   d. Construct

   e. Content

17. _____ validity examines how well a test reflects an underlying idea.

   a. Predictive

   b. Concurrent criterion

   c. Criterion

   d. Construct

   e. Content

18. _____ validity examines the extent to which a test accurately reflects all possible topics and ideas under the subject or topic it is designed to test for.

   a. Predictive

   b. Concurrent criterion

   c. Criterion

   d. Construct

   e. Content

19. If you need to improve content validity, you should do which of the following?

    a. Reexamine the nature of the items on the test and question whether responses should be expected to relate to the criterion you selected.

    b. Review the theoretical rationale that underlies the test.

    c. Redo questions so that, according to an expert judge, they more accurately reflect the topics and ideas being tested for.

20. If you need to improve criterion validity, you should do which of the following?

    a. Reexamine the nature of the items on the test and question whether responses should be expected to relate to the criterion you selected.

    b. Review the theoretical rationale that underlies the test.

    c. Redo questions so that, according to an expert judge, they more accurately reflect the topics and ideas being tested for.

21. If you need to improve construct validity, you should do which of the following?

    a. Reexamine the nature of the items on the test and question whether responses should be expected to relate to the criterion you selected.

    b. Review the theoretical rationale that underlies the test.

    c. Redo questions so that, according to an expert judge, they more accurately reflect the topics and ideas being tested for.

22. Which of the following is a correct statement?

    a. The maximum reliability is equal to the square root of the validity.

    b. The maximum validity is equal to the square root of the reliability.

    c. The maximum reliability is equal to the square of the validity.

    d. The maximum validity is equal to the square of the reliability.

23. In order to measure attitudes about women on a newly developed scale, a researcher compares scores on his new scale with other scales that focus on attitudes toward women (like the Ambivalent Sexism Inventory). If they have similar results, what kind of validity has the research established?

    a. Predictive validity

    b. Concurrent validity

    c. Content validity

    d. Construct validity

24. Bethany and Paul are studying cuddling behavior among couples attending romantic movies. They both attend the movies and count the number of times couples in the audience cuddle, and then they compare their scores. Bethany and Paul are using which kind of reliability in their score comparison?

    a. Parallel forms reliability

    b. Test-retest reliability

    c. Internal consistency reliability

    d. Interrater reliability

25. Roger develops a new scale to measure beliefs about acceptance of illegal immigrants in the United States. She gives her new scale to 100 people twice, two months apart. She finds that in general, people who are more accepting of illegal immigrants at time 1 are also more accepting of illegal immigrants at time 2. Roger's scale exhibits good _____.

    a. parallel forms reliability

    b. test-retest reliability

    c. internal consistency reliability

    d. interrater reliability

26. The Meyers-Briggs inventory was designed to indicate psychological preferences among people (whether their primary perception of the world is based on thinking, feeling, sensation, or intuition). However, research indicates that between 39% and 76% of respondents obtain a different classification when retaking the inventory five months later. This indicates low _____.

    a. parallel forms reliability

    b. test-retest reliability

    c. internal consistency reliability

    d. interrater reliability

EXERCISES

1. Say that you were very tired when taking an exam and did much worse than you had anticipated, getting a score of 72. If you had been wide awake and energetic, and if the test perfectly measured your knowledge, you would have gotten a score of 94. What is your error score?

2. The following data illustrate a set of scores from an instrument administered to eight respondents at two different times. Calculate the test-retest reliability.

ID	Time 1	Time 2
1	72	76
2	54	43
3	86	92
4	92	94
5	99	87
6	43	69
7	76	43
8	84	92

3. The following data illustrate a set of scores from a test administered in two different forms to five respondents. Calculate the parallel forms reliability.

ID	Form 1	Form 2
1	8.7	9.2
2	7.6	7.9
3	3.4	4.3
4	3.7	5.3
5	5.4	6.1

4. Two psychologists view a clinical session involving a child known for slapping his head whenever he gets frustrated. They watch the child try to accomplish a frustrating task across ten trials and count whether he does or does not slap himself during each trial. The following data illustrate whether each psychologist thought slapping was present (P) or absent (A). Calculate the interrater reliability.

Rater	1	2	3	4	5	6	7	8	9	10
Rater #1	P	A	P	P	P	A	P	P	A	P
Rater #2	P	P	A	P	P	A	P	P	A	A

SHORT-ANSWER/ESSAY QUESTIONS

1. Explain the difference between test-retest reliability and parallel forms reliability.

2. If you don't have validity, then your test probably isn't doing what it should be doing. What can you do to increase validity?

3. Compare and contrast reliability and validity. Define those concepts first, and then note how they interact with one another. Can you have one without the other? Under what circumstances could something be valid but not reliable, or reliable but not valid?

SPSS QUESTIONS

1. Input the data from question 1 under the "Just for Fun/Challenge Yourself" section below into SPSS and calculate Cronbach's alpha. If you complete this "Challenge Yourself" question manually, do the two figures match?

JUST FOR FUN/CHALLENGE YOURSELF

1. The following data present scores for 10 individuals on a 5-item test. Using these data, calculate Cronbach's alpha. Now, do a little additional research on this statistic. Is the score you calculated good or bad? Is it high enough to be considered acceptable?

ID	Item 1	Item 2	Item 3	Item 4	Item 5
1	8	9	7	8	8
2	5	6	8	4	5
3	1	2	1	1	1
4	4	3	4	5	4
5	1	3	2	3	2
6	5	4	5	6	5
7	2	3	4	2	3
8	9	7	8	7	9
9	6	5	8	5	6
10	4	5	5	4	5

ANSWER KEY

TRUE/FALSE QUESTIONS

1. False. Reliability and validity should always be examined before an analysis is conducted. If the reliability and/or validity is not up to par, you risk having inconclusive results.

2. True. Error is measured as the difference between the true score and the observed (measured) score. If these two scores were found to be exactly the same, you would have no error, or perfect reliability.

3. False. This is extremely rare in the social and behavioral sciences due to the common presence of error.

4. True. The less error you have, the greater your reliability.

5. False. High reliability is in fact associated with high reliability coefficients, as close as possible to 1.

6. False. Having very low reliability is a serious concern. The steps appropriate for the type of reliability in question should be taken to attempt to increase the level of reliability.

7. False. Instead, it is best to use a previously published test instrument whose reliability and validity have already been determined to be at least adequate. Otherwise, you run the serious and substantial risk of having your test instrument lack sufficient reliability and/or validity, which would call your data into serious question and force your results to be inconclusive.

8. True. Reliability coefficients range from .00 to +1.00. A Chronbach's alpha of .811 is very high on that scale, showing that the happiness questions are related.

9. False. Interrater reliability involves looking at the number of agreements divided by the possible number of agreements. Since the two raters only agreed 7 out of 12 times, total interrater reliability is .58.

10. False. Reliability must show consistency while validity must show that the tool does what it says it does. If a tool finds something different each time, it is neither reliable (consistent) nor doing what it should be doing (valid). Yet a tool can be reliable but not valid.

MULTIPLE-CHOICE QUESTIONS

1. (b) the independent

2. (a) the dependent

3. (c) How do I know that the test, scale, instrument, etc., I use works every time I use it?

4. (c) How do I know that the test, scale, instrument, etc., I use measures what it is supposed to?

5. (b) observed

6. (a) true

7. (d) error

8. (d) Consistency over time

9. (b) The consistency across different forms of the same test

10. (c) The extent to which items in a test are consistent with one another and represent exactly one dimension, construct, or area of interest

11. (a) The consistency of judges

12. (c) Internal consistency

13. (d) Interrater

14. (a) Test-retest

15. (b) Parallel forms

16. (c) Criterion

17. (d) Construct

18. (e) Content

19. (c) Redo questions so that, according to an expert judge, they more accurately reflect the topics and ideas being tested for.

20. (a) Reexamine the nature of the items on the test and question whether responses should be expected to relate to the criterion you selected.

21. (b) Review the theoretical rationale that underlies the test you developed.

22. (b) The maximum validity is equal to the square root of the reliability.

23. (b) Concurrent validity

24. (d) Interrater reliability

25. (b) test-retest reliability

26. (b) test-retest reliability

EXERCISES

1. Error score = Observed score − True score = 72 − 94 = −22.

2. The necessary calculations for this question are shown here. In essence, this question necessitates calculation of the correlation between time 1 scores and time 2 scores.

ID	Time 1	Time 2	Time 1²	Time 2²	Time 1 * Time 2
1	72	76	5,184	5,776	5,472
2	54	43	2,916	1,849	2,322
3	86	92	7,396	8,464	7,912
4	92	94	8,464	8,836	8,648
5	99	87	9,801	7,569	8,613
6	43	69	1,849	4,761	2,967
7	76	43	5,776	1,849	3,268
8	84	92	7,056	8,464	7,728
Sum:	606	596	48,442	47,568	46,930

$$r_{xy} = \frac{n\Sigma XY - \Sigma X \Sigma Y}{\sqrt{(n\Sigma X^2 - (\Sigma X)^2)(n\Sigma Y^2 - (\Sigma Y)^2)}}$$

$$= \frac{8 \times 46,930 - 606 \times 596}{\sqrt{(8 \times 48,442 - 606^2)(8 \times 47,568 - 596^2)}}$$

$$= \frac{375,440 - 361,176}{\sqrt{(20,300)(25,328)}}$$

$$= .6291$$

3. The following illustrates the calculations necessary to calculate parallel forms reliability. In essence, this question requires the calculation of the correlation between form 1 scores and form 2 scores.

ID	Form 1	Form 2	Form 1²	Form 2²	Form 1 * Form 2
1	8.7	9.2	75.69	84.64	80.04
2	7.6	7.9	57.76	62.41	60.04
3	3.4	4.3	11.56	18.49	14.62
4	3.7	5.3	7.29	28.09	14.31
5	5.4	6.1	29.16	37.21	32.94
Sum:	27.8	32.8	181.46	230.84	201.95

$$r_{xy} = \frac{n\Sigma XY - \Sigma X \Sigma Y}{\sqrt{(n\Sigma X^2 - (\Sigma X)^2)(n\Sigma Y^2 - (\Sigma Y)^2)}}$$

$$= \frac{5 \times 201.95 - 27.8 \times 32.8}{\sqrt{(5 \times 181.46 - 27.8^2)(5 \times 230.84 - 32.8^2)}}$$

$$= \frac{1,009.75 - 911.84}{\sqrt{(134.46)(78.36)}}$$

$$= .9539$$

4. The following illustrates the calculation of the interrater reliability for this question: Divide the number of total agreements by the number of total possible agreements.

Rater	1	2	3	4	5	6	7	8	9	10
Rater #1	P	A	P	P	P	A	P	P	A	P
Rater #2	P	P	A	P	P	A	P	P	A	A

$$\text{Interrater reliability} = \frac{n \text{ Agreements}}{n \text{ Possible agreements}} = \frac{7}{10} = .7$$

SHORT-ANSWER/ESSAY QUESTIONS

1. Test-retest reliability measures reliability, or consistency over time, while parallel forms reliability examines consistency across multiple forms of the same test. Test-retest reliability would be calculated by determining the correlation between a test given at one time and the same test given at a second time, while parallel forms reliability would be calculated by determining the correlation between administrations of two versions of the test instrument.

2. You can do several things to increase validity. First, rewrite some questions to achieve better consistency with the construct. Second, seek out expert opinions for the construct to make sure you are actually measuring what you think you are measuring. Third, examine the theoretical rationale for the constructs in the survey.

3. Validity refers to whether a survey is measuring what it intends to measure. Reliability refers to the extent to which a test or inventory is consistent in evaluating the constructs (repeatable). Although researchers want both to be high, that isn't always possible. Reliability is the easy one. If you can find something consistently, it is reliable, but that doesn't mean it is valid. I can call a circle a square a thousand times (and be consistent each time I call it a square), but that doesn't mean it is valid. A valid score MUST be reliable (or you aren't measuring it correctly).

SPSS QUESTIONS

The SPSS output for this analysis is presented below. The calculated Cronbach's alpha score is 0.969, which is identical to the figure you would have arrived at by calculating manually.

Case Processing Summary

		N	%
Cases	Valid	10	100.0
	Excluded[a]	0	.0
	Total	10	100.0

[a] Listwise deletion based on all variables in the procedure.

Reliability Statistics

Cronbach's Alpha	N of Items
.969	5

JUST FOR FUN/CHALLENGE YOURSELF

1. First, the variances for all five items as well as the variance for total score need to be calculated. As this calculation was covered in Chapter 3, it is not repeated here. The final column of the following table gives the variances of each of the five individual items as well as of the total score.

ID	Item 1	Item 2	Item 3	Item 4	Item 5	Total
1	8	9	7	8	8	40
2	5	6	8	4	5	28
3	1	2	1	1	1	6
4	4	3	4	5	4	20
5	1	3	2	3	2	11
6	5	4	5	6	5	25
7	2	3	4	2	3	14
8	9	7	8	7	9	40
9	6	5	8	5	6	30
10	4	5	5	4	5	23
Variance:	7.39	4.68	6.40	4.72	6.18	130.46

Next, the variances for the five individual items need to be summed:

Sum of item variances = 29.37.

Finally, the following equation presents the calculation for Cronbach's alpha:

$$\alpha = \left(\frac{k}{k-1}\right)\left(\frac{s_Y^2 - \Sigma s_i^2}{s_Y^2}\right)$$

$$\Rightarrow \left(\frac{5}{5-1}\right)\left(\frac{130.46 - 29.37}{130.46}\right)$$

$$= .9686$$

These five items have a Cronbach's alpha score of 0.9686. Generally, alpha scores of 0.7 or higher are considered acceptable. Thus, these five items have a very high and very acceptable alpha score.

7 Hypotheticals and You

Testing Your Questions

LEARNING OBJECTIVES

- Learn the differences between a sample and a population and how each is related to your research.

- Understand the difference between null and research hypotheses, as well as directional and nondirectional hypotheses.

- Learn how to create good hypotheses.

SUMMARY/KEY POINTS

- A hypothesis is an "educated guess" that describes the relationship between variables. In essence, it is like a more specific, directly testable version of a research question.

- In a study, a sample is drawn from a larger population, and analyses are conducted on the sample. Optimally, it is possible to generalize your results to the population.
  - Hypothesis testing relates to the sample itself, not the population.
  - A *representative* sample must be used if you wish to generalize the results of your analyses to the population at large. Individuals in a representative sample should match as closely as possible to the characteristics of the population (and the study must also meet more specific methodological requirements). If you want to assess the effectiveness of a new program designed to decrease alcohol consumption on college campuses and you only look at nontraditional, older students, the study will not fully represent all students on campus.

- Sampling error is how well a sample approximates the characteristics of a population.
  - Higher sampling error means a greater difference between the sample statistic and the population parameter, so it is more difficult to generalize your results to the population. If you recruit participants for a study in Miami and find that 80% of your sample is Caucasian, you will have a difficult time generalizing your results from this sample to a Miami population (where the Hispanic/Latino population is actually 70%).

- The two types of hypotheses are the null and alternative hypothesis.
  - The null hypothesis is formulated first and states that there is no relationship between your variables.
  - The research hypothesis, formulated second, states that there is a relationship between your variables.
    - A research hypothesis that suggests the direction of the relationship is called a *directional* hypothesis. One-tailed tests can be used with these hypotheses. For example, you might predict that a new social program designed to train teens to combat peer pressure might lower their likelihood to yield to peer pressure. Here, you do not care whether the new program leads teens to yield more often. You are only interested in whether they yield less often.
    - A research hypothesis that does not suggest the direction of the relationship is called a nondirectional hypothesis. Two-tailed tests should be used with these hypotheses. Here, you might be interested in determining whether the new program leads to more or less likelihood of yielding to peer pressure. This allows the researcher to also determine whether the new program leads to less resistance. Sometimes it is important for the researcher to know both the positive and negative consequences of a program.

○ Unless you have sufficient evidence otherwise, you must assume that the null hypothesis is true. That is, you assume the new peer pressure social program is ineffective.
○ Null hypotheses always refer to the population, while research hypotheses always refer to the sample. Therefore, the null hypothesis is only indirectly tested (making it an *implied* hypothesis), while the research hypothesis can be tested directly.
○ While null hypotheses are written using Greek symbols, research hypotheses are written using Roman symbols (letters from the English alphabet).
○ Good hypotheses have the following features:
  ○ They are stated in declarative form, not as a question.
  ○ They posit an expected relationship between variables.
  ○ They reflect the theory or literature on which they are based.
  ○ They are brief and to the point.
  ○ They are testable.
  ○ A good research hypothesis for the peer pressure social program might be, "Teenagers who are trained to resist peer pressure using the new social program will be better able to resist peer pressure compared to teenagers who are not trained to resist peer pressure." The null hypothesis would note that training does not impact peer pressure resistance.

KEY TERMS

- **Hypothesis:** An "educated guess" describing the relationship between two or more variables

- **Population:** A larger group of respondents from whom a sample is collected and to which you hope to generalize after conducting your analyses

- **Sample:** A subset taken from a population for the purposes of your study. Data are collected on a sample, and, optimally, the results are generalized to the population.

- **Sampling error:** The difference between sample and population values

- **Null hypothesis:** A statement of equality between sets of variables

- **Research hypothesis:** A statement that there is a relationship between variables

- **Nondirectional research hypothesis:** A hypothesis that reflects a difference between groups but does not specify the direction of the difference

- **Directional research hypothesis:** A hypothesis that reflects a difference between groups and also specifies the direction of the difference

- **One-tailed test:** A directional test, which reflects a directional hypothesis

- **Two-tailed test:** A nondirectional test, which reflects a nondirectional hypothesis

TRUE/FALSE QUESTIONS

1. A research question is a more specific, testable version of a hypothesis.

2. The null hypothesis is only indirectly tested, making it an implied hypothesis.

3. Good hypotheses should be (along with other attributes) brief and to the point, testable, and stated in declarative form.

4. "If you give participants a new drug designed to combat social anxiety, they will have less anxiety" represents a nondirectional hypothesis.

5. A researcher predicts that distracting noise will lower the accuracy of participants taking an exam. The researcher should use a one-tailed test when analyzing the results of the study.

6. A researcher studies views about healthcare by posting a pop-up survey online for visitors to the government healthcare website. He gets a response rate of around 30% from a population of 12,000 potential participants, with most of the responding sample under the age of 25. He probably has a high level of sampling error in this study.

MULTIPLE-CHOICE QUESTIONS

1. When conducting a study, you draw a smaller _____ from a larger _____.

   a. population; sample

   b. sample; population

   c. null hypothesis; research hypothesis

   d. nondirectional hypothesis; directional hypothesis

2. To generalize your results, you need to have a _____.

   a. representative sample

   b. representative population

   c. population larger than your sample

   d. null hypothesis

3. A representative sample should have which of the following?

   a. A null hypothesis

   b. A nondirectional hypothesis

   c. A population

   d. A small level of sampling error

4. A high level of sampling error means that _____.

   a. the population may be too small

   b. your sample may be too large

   c. you may not be able to generalize to your population

   d. your hypotheses will not be supported

5. Which of the following types of hypotheses states that there is no relationship between your variables?

   a. The null hypothesis

   b. The research hypothesis

   c. The population hypothesis

   d. The sample hypothesis

6. Which of the following types of hypotheses states that there is a relationship between your variables?

	a. The null hypothesis

	b. The research hypothesis

	c. The population hypothesis

	d. The sample hypothesis

7. A one-tailed test would be used with a _____ hypothesis.

	a. null

	b. research

	c. directional

	d. nondirectional

8. A two-tailed test would be used with a _____ hypothesis.

	a. null

	b. research

	c. directional

	d. nondirectional

9. If you are unsure whether the null or research hypothesis is true, you must assume that _____.

	a. the population was too small

	b. the sample was too small

	c. the research hypothesis is true

	d. the null hypothesis is true

10. Null hypotheses always refer to _____.

	a. the sample

	b. the population

	c. sampling error

	d. two-tailed tests

11. Research hypotheses always refer to _____.

	a. the sample

	b. the population

	c. sampling error

	d. one-tailed tests

12. Null hypotheses are written using which of the following types of letters?

    a. Greek

    b. Roman

    c. Arabic

    d. Sanskrit

13. Research hypotheses are written using which of the following types of letters?

    a. Greek

    b. Roman

    c. Arabic

    d. Sanskrit

14. Which of the following is not a feature of good hypotheses?

    a. There should be no more than one null and one research hypothesis in any study.

    b. They should be brief and to the point.

    c. They should be testable.

    d. They should posit an expected relationship between variables.

15. Which type of hypothesis is this? *There is no relationship between religiosity and drug use.*

    a. Research hypothesis

    b. Null hypothesis

    c. Sample hypothesis

    d. Directional research hypothesis

16. Which type of hypothesis is this? *Individuals with a college degree will have higher incomes than those with no college degree.*

    a. Null hypothesis

    b. Directional research hypothesis

    c. Nondirectional research hypothesis

    d. Sample hypothesis

17. Which type of hypothesis is this? *Group A will differ from Group B with regard to test scores.*

    a. Null hypothesis

    b. Directional research hypothesis

    c. Nondirectional research hypothesis

    d. Sample hypothesis

18. What type of hypothesis is this an example of? $H_1 : \bar{X}_A \neq \bar{X}_B$.

    a. Null hypothesis

    b. Directional research hypothesis

    c. Nondirectional research hypothesis

    d. Population hypothesis

19. What type of hypothesis is this an example of? $H_1 : \bar{X}_A < \bar{X}_B$.

    a. Null hypothesis

    b. Directional research hypothesis

    c. Nondirectional research hypothesis

    d. Population hypothesis

20. What type of hypothesis is this an example of? $H_0 : \mu_A = \mu_B$.

    a. Null hypothesis

    b. Directional research hypothesis

    c. Nondirectional research hypothesis

    d. Population hypothesis

21. "Depressed individuals who receive therapy will be less depressed than those in the control group who do not receive therapy." This hypothesis is an example of a _____.

    a. contradictory hypothesis

    b. nondirectional research hypothesis

    c. directional research hypothesis

    d. nonfalsifiable hypothesis

22. "Children who play video games will differ in their hand-eye coordination from children who do not play video games." This hypothesis is an example of a _____.

    a. contradictory hypothesis

    b. nondirectional research hypothesis

    c. directional research hypothesis

    d. nonfalsifiable hypothesis

23. Which of the following best represents a hypothesis involving emotion?

    a. A person can experience physiological arousal first, and then interpret those feelings as arousal.

    b. A person can experience arousal and interpret those feelings as arousal at the same time.

    c. An event causes arousal first, and then the person has to label this arousal.

    d. Exposing people to torture videos (compared to videos about nature) will increase their feelings of arousal (as measured by heart rates and breathing rates).

24. Which of the following is a good example of a null hypothesis?

    a. Participants who receive a high score (versus a low score) on an exam will have higher ratings of self-esteem.

    b. Those who take a new GRE prep program will score differently than those who do not take the GRE prep program.

    c. Participants who receive positive feedback from a teacher will not differ from those who receive negative feedback.

    d. Giving students a hard task will decrease their performance in comparison to those given an easy task.

SHORT-ANSWER/ESSAY QUESTIONS

1. You are going to conduct a psychological study focusing on the relationship between watching violence on television and violent behavior. Generate a null hypothesis and two research hypotheses (nondirectional and directional) for this study.

2. Now, write both the null and research hypotheses using the appropriate Greek or Roman letters.

3. Check the hypotheses you generated against the five features of a good hypothesis. Do your hypotheses fulfill all of the requirements?

4. Come up with written descriptions of the hypotheses given under questions 18–20 in the "Multiple-Choice" section.

5. What is wrong with this hypothesis? *Will the accelerated class score higher on a reading comprehension exam as compared with a regular class?*

6. What is wrong with this hypothesis? *Religious leaders during the Stone Age gave greater thought to the meaning of life than did religious leaders at other times.*

7. Under what circumstances is it best to use a nondirectional hypothesis rather than a directional hypothesis?

8. Some scholars argue that jurors who watch a lot of crime TV come to expect CSI-based high-tech evidence at trials, though high-tech evidence does not always occur in real courtrooms. If jurors do not get this evidence, the "CSI Effect" theory states that such jurors will find less evidence of guilt. Develop a directional and a nondirectional hypothesis to assess this theory.

9. Since all studies have some degree of sampling error, why not just study the whole population?

10. There are times when researchers would prefer to retain the null hypothesis and reject the research hypothesis. Design a study in which the researcher would want to support the null hypothesis. Argue for why the study should use a directional or nondirectional approach.

JUST FOR FUN/CHALLENGE YOURSELF

Do some additional research on the difference between one-way and two-way statistical tests.

1. How do they differ?

2. Which of the two is more likely to be found significant, in general?

3. Why is it okay to use a one-way statistical test with a directional hypothesis but not with a nondirectional hypothesis?

ANSWER KEY

TRUE/FALSE QUESTIONS

1. False. The opposite is true—a hypothesis is a more specific, testable version of a research question.

2. True.

3. True. These are some of the characteristics of good hypotheses.

4. False. This represents a directional hypothesis. A nondirectional hypothesis would be, "If you give participants a new drug to combat social anxiety, they will differ from those not taking the new social anxiety drug."

5. True. A one-tailed test assesses whether a treatment does better than an alternative with no need to determine if the treatment does better *or* worse.

6. True. Internet studies might draw a younger, more computer-literate crowd, which may not adequately reflect the larger population.

MULTIPLE-CHOICE QUESTIONS

1. (b) sample; population

2. (a) representative sample

3. (d) A small level of sampling error

4. (c) you may not be able to generalize to your population

5. (a) The null hypothesis

6. (b) The research hypothesis

7. (c) directional

8. (d) nondirectional

9. (d) the null hypothesis is true

10. (b) the population

11. (a) the sample

12. (a) Greek

13. (b) Roman

14. (a) There should be no more than one null and one research hypothesis in any study.

15. (b) Null hypothesis

16. (b) Directional research hypothesis

17. (c) Nondirectional research hypothesis

18. (c) Nondirectional research hypothesis

19. (b) Directional research hypothesis

20. (a) Null hypothesis

21. (c) directional research hypothesis

22. (b) nondirectional research hypothesis

23. (d) Exposing people to torture videos (compared to videos about nature) will increase their feelings of arousal (as measured by heart rates and breathing rates). Note that the first three options are actually *theories* of emotion, including the Schachter-Singer Theory, Cannon-Bard Theory, and James Lange Theory, respectively.

24. (c) Participants who receive positive feedback from a teacher will not differ from those who receive negative feedback.

SHORT-ANSWER/ESSAY QUESTIONS

1. An example of a null hypothesis: *With regard to the number of instances of violent behavior, there will be no relationship between individuals who watch less than 1 hour of violent television per day and those who watch 1 hour or more of violent television per day.*

   An example of a nondirectional research hypothesis: *With regard to the number of instances of violent behavior, there is a difference between individuals who watch less than 1 hour of violent television per day and those who watch 1 hour or more of violent television per day.*

   An example of a directional research hypothesis: *Individuals who watch 1 hour of violent television per day or more will exhibit a greater number of instances of violent behavior than those who watch less than 1 hour of violent television per day.*

2. Here are some examples (your subscripts do not have to match exactly):

   Null hypothesis: $H_0 : \mu_{<1hr} = \mu_{1+hr}$.

   Nondirectional research hypothesis: $H_1 : \bar{X}_{<1hr} \neq \bar{X}_{1+hr}$.

   Directional research hypothesis: $H_1 : \bar{X}_{<1hr} < \bar{X}_{1+hr}$.

3. As a reminder, the five features are these:
   ○ They are stated in declarative form, not as a question.
   ○ They posit an expected relationship between variables.
   ○ They reflect the theory or literature on which they are based.
   ○ They are brief and to the point.
   ○ They are testable.

4.

   a. For question 18: $H_1 : \bar{X}_A \neq \bar{X}_B$.

      The average score of individuals in Group A is different from the average score of individuals in Group B.

For question 19: $H_1 : \bar{X}_A < \bar{X}_B$.

Individuals in Group B will have higher scores, on average, than will individuals in Group A.

For question 20: $H_0 : \mu_A = \mu_B$.

There is no difference between the average score of individuals in Group A and the average score of individuals in Group B.

5. The problem with this hypothesis is that it is phrased as a question.

6. The problem with this hypothesis is that it is not testable (we don't have the data and can't collect it).

7. If a researcher thinks the results might turn out opposite of what they predicted, they should use a nondirectional approach. If they only want to see whether a new treatment helps participants, a directional approach would be acceptable. However, if they think the treatment might harm patients, they should use a nondirectional approach.

8. A directional hypothesis might posit, "Jurors who are given high-tech CSI evidence will find more evidence of guilt than jurors not given such evidence." A nondirectional hypothesis might posit, "Jurors who are given high-tech CSI evidence will differ in their assessments of guilt from jurors not given such evidence."

9. It is often impossible to study the whole population, as it is costly, time-consuming, and bound to have error of its own. For example, the United States constitution requires that the government conduct a population census every ten years to count all US citizens. Yet the census still misses a lot of people. People may be on vacation or away from home when census takers appear, or they might be homeless or refuse to answer census questions.

10. A researcher may want to see if implementing a new, low-cost training program would be just as good as a current, high-cost training program. In this case, the researcher would use a nondirectional approach to assess the differences between the two programs to determine whether the new program was better or worse than the old program. She would hope that the new and current programs would not differ in terms of effectiveness (retain the null hypothesis), which would argue for adopting the new program with its lower cost.

JUST FOR FUN/CHALLENGE YOURSELF

1. As mentioned in this chapter, one-way statistical tests are suited to directional hypotheses, while two-way tests are suited to nondirectional hypotheses. In essence, a two-way test examines both possibilities (i.e., that group 1 has a higher average than group 2 and that group 2 has a higher average than group 1). A one-way test examines only one possibility (i.e., that either group 1 has a higher average than group 2 or that group 2 has a higher average than group 1).

2. Like a directional hypothesis, a one-way statistical test measures only one outcome (Group A is better than Group B). A directional hypothesis requires testing more than one outcome (Group A is better OR worse than Group B).

3. It is not justified to use a one-way test with a nondirectional hypothesis, as you are going into the analysis not knowing what you may find. Therefore, using a more powerful one-way test in situations where you don't initially start with a directional hypothesis is, in a sense, "cheating."

8

Are Your Curves Normal?

Probability and Why It Counts

LEARNING OBJECTIVES

- Review the importance of probability in statistics.

- Understand the normal curve and its relation to the field of statistics.

- Learn how to compute z scores by hand and using SPSS.

SUMMARY/KEY POINTS

- The study of probability is the basis for the normal curve and the foundation for inferential statistics.
  - The normal curve provides a basis for understanding the probability associated with any possible outcome, such as attaining a certain score.
  - The study of probability is the basis for determining the degree of confidence we have in stating that a particular finding or outcome is true.
  - Probability allows us to determine the exact mathematical likelihood that a difference between groups, or an association between variables, is due to a practice or treatment rather than to chance or error. For example, a researcher might want to see how likely someone in need will get assistance depending on whether a potential helper is alone (and is thus the only possible helper) or in a group of six people (where several people can take on the role of helper). The researcher could assess the probability of help occurring in the "alone" condition compared to the "group" condition.

- The normal curve is the basis for probability and statistics.
  - The normal curve has no skew and is perfectly symmetrical about the mean.
  - The tails of the normal distribution are asymptotic, meaning they never touch the horizontal axis, which is equal to zero.
  - In the social and behavioral sciences, as well as in other fields, many things are normally distributed, including measures such as height and IQ. The average IQ is 100, with most people scoring 100. While there are some people with an IQ of 110 and some with an IQ of 90, there are fewer such individuals. Even rarer are those with an IQ of 120 or an IQ of 80.
  - Events that occur in the extremes of the normal curve have a very small probability, while more "average" values are much more common. The small number of people invited to join MENSA must have an IQ of at least 135.
  - In a normal curve, the mean, median, and mode all line up in the middle of the curve.

- The normal curve has many specific statistical features.
  - Over 99.5% of scores are within 3 standard deviations of the mean.
  - Approximately 68% of scores fall within 1 standard deviation of the mean. The average height of men is 70 inches, with 34% falling between 70 inches and 73 inches and 34% falling between 67 inches and 70 inches. Thus, 68% of men fall between 67 and 73 inches tall.
  - Exactly 50% of scores fall on either side of the distribution (i.e., either side of the mean).
  - The percentages or areas under the normal curve can be interpreted as probabilities.

- Standard scores are raw scores that have been adjusted for the particular mean and standard deviation of the distribution from which they are derived. They can be used to compare raw scores between different samples that have different distributions. Thus, the standard score for the height of men is comparable to the standard score for the height of women, even though men are taller on average than women.

The most commonly used standard score is the z score.
- ○ To calculate the z score, you subtract the mean from the raw score and divide this difference by the standard deviation.
- ○ Scores that fall below the mean have negative z scores, while scores that fall above the mean have positive z scores.
- ○ The score located 1 standard deviation above the mean is "1 z score" above the mean.
- ○ We can use z scores and the normal distribution to determine the probability of some event occurring.
- ○ Skewness and kurtosis are measures that are used to describe the shape of a distribution.

- A statistical test can be used to determine the probability of the differences between groups or relationships between variables in the data. After the test is conducted, the calculated probability can be compared with a standard to see whether the result is "significant."

The standard of 5%, which is equivalent to a probability of .05, is the most commonly used standard in statistics. This means we need to be at least 95% sure of the difference between groups or the relationship between variables in order to call it "significant." This also means that a result is significant if we find a z score that has less than a 5% chance of occurring.

KEY TERMS

- **Normal curve** (bell-shaped curve): A distribution of scores that is symmetrical about the mean and in which the median, mean, and mode are all equal. This type of distribution has asymptotic tails, which never reach zero.

- **Asymptotic**: The quality of the normal curve such that its tails never touch the horizontal axis (equal to zero)

- **Standard scores**: Raw scores that are adjusted for the mean and standard deviation of the distribution from which they come

- **Standardized scores**: A score that comes from a distribution with a predefined mean and standard deviation

- *z* **score**: A specific type of standard score in which the mean of scores is subtracted from the raw score and then this difference is divided by the standard deviation
  - ○ **Skewness**: A measure of the lack of symmetry, or "lopsidedness," of a distribution. A distribution that is skewed has one tail that is longer than the other.
    - ○ **Positive skew**: A distribution that has many data points to the left and a long tail to the right
    - ○ **Negative skew**: A distribution that has many data points to the right and a long tail to the left
  - ○ **Kurtosis**: A measure that relates to how flat or peaked a distribution appears
    - ○ **Platykurtic**: A distribution that is relatively flat compared to a normal, or bell-shaped, distribution
    - ○ **Leptokurtic**: A distribution that is relatively peaked compared to a normal, or bell-shaped, distribution

TRUE/FALSE QUESTIONS

1. The percentages of scores under sections of the normal curve depend on the mean and standard deviation of distribution.

2. You can compare z scores across two or more different distributions.

3. Values for the area under the normal curve can be viewed/interpreted as probabilities.

4. When looking up z scores using a z table, it is very important to consider whether the z score you are looking up is positive or negative.

5. A standard score is the same as a standardized score; an example is the score you received on the SAT or ACT.

6. Skewness is a measure of the central point of a class interval.

7. A psychologist measures satisfaction in personal relationships among couples in marriage counseling using a scale ranging from 0 to 20, with higher scores representing more satisfaction. The mean score is 10 and the standard deviation is 3. For a raw score of 3, the resulting z score is -2.33.

8. A researcher gives a gifted student a cognitive test to assess the student's academic potential. The student scores 148 (mean 125, standard deviation 15). The percentile rank for the student is 43.7.

9. Roger and James take a test on spatial ability (mean 80, standard deviation 10). Roger scores 70 and James scores 92. The percent of individuals who score between Roger and James is 65.14%.

10. A normal curve will eventually touch the x-axis.

MULTIPLE-CHOICE QUESTIONS

1. Which of the following percentages of the normal curve reflects all scores greater than zero?

   a. 10%

   b. 25%

   c. 50%

   d. 100%

2. The entire normal curve represents _____ of scores.

   a. 25%

   b. 50%

   c. 99%

   d. 100%

3. With regard to the normal curve, _____ of scores are within 1 standard deviation of the mean.

    a. 13.59%

    b. 2.15%

    c. 68.26%

    d. 99.99%

4. Which of the following percentages of scores is within 2 standard deviations of the mean?

    a. 0.13%

    b. 2.15%

    c. 68.26%

    d. 95.44%

5. If your set of scores has a mean of 57, what is the z score for a raw score of 57?

    a. −1

    b. 0

    c. 1

    d. 2

6. If a raw score is above the mean, the z score must be _____.

    a. negative

    b. positive

    c. equal to zero

    d. impossible to compute

7. If a score is 4 standard deviations above the mean, then its z score must be equal to _____.

    a. −4

    b. 4

    c. 0

    d. 1

    e. 4^+

8. Based on the normal curve, what percentage of scores have a z score of 2 or greater?

    a. 7.16%

    b. 22.42%

    c. 2.28%

    d. 0.15%

9. Based on the normal curve, what percentage of scores have a z score less than 20.5?

    a. 30.85%

    b. 22.42%

    c. 47.40%

    d. 23.65%

10. On your last exam, the class scored an average of 82 with a standard deviation of 8. What's the probability of any one student's score being 90 or greater?

    a. 12.15%

    b. 17.31%

    c. 23.52%

    d. 15.87%

11. Using the same scenario as in question 10, what's the probability of any one student's score being a failing grade (65 or less)?

    a. 1.68%

    b. 2.42%

    c. 12.40%

    d. 7.14%

12. On your last exam, the class scored an average of 72 with a standard deviation of 12. What is the probability of any one student's z score being between 70 and 80?

    a. 31.61%

    b. 21.42%

    c. 17.67%

    d. 12.14%

13. Which of the following is the most common probability standard used by researchers when conducting analyses?

    a. .01

    b. .05

    c. .10

    d. .50

14. This is defined as "a measure of the lack of symmetry, or 'lopsidedness,' of a distribution."

    a. Skewness

    b. Kurtosis

    c. Mean

    d. Variance

    e. Median

15. This is defined as "a measure that relates to how flat or peaked a distribution appears."

    a. Skewness

    b. Kurtosis

    c. Variance

    d. Mode

    e. Leptokurtic

16. A distribution that is relatively flat compared to a normal distribution is called _____.

    a. skewed

    b. platykurtic

    c. leptokurtic

    d. mesokurtic

17. A sports psychologist is interested in a school physical fitness program. He knows that the mean time to run a mile among a normal distribution of students is 440 seconds with a standard deviation of 60 seconds. What is the probability of him finding a randomly selected boy who can run the mile in less than 302 seconds?

    a. .9893

    b. .0107

    c. .5107

    d. .4893

18. On a test of emotional intelligence, a participant has a z score of .44. How many participants would score above her?

    a. 33%

    b. 53%

    c. 73%

    d. 83%

EXERCISES

1. Your set of scores has a mean of 5.8 and a standard deviation of 2.3. Calculate the z scores for the following raw scores: 2.1, 5.7, 7.3, and 12.4.

2. You are taking a standardized test, and you want to score in the top 5% of test takers. You know that the mean is 1,000 and the standard deviation is 100. What is the minimum score you need in order to rank in the top 5%?

3. A set of test scores has a mean of 27 and a standard deviation of 4.2. Calculate the raw scores for each of the following z scores: −5.3, −2.1, 0, 1, and 3.1.

SHORT-ANSWER/ESSAY QUESTIONS

1. What are some examples of measures that you think may be normally distributed? In general, what does the distribution of a measure need to look like for it to be normally distributed?

2. While counseling Jerry, a child with attention problems in school, you reach out to his instructors to find out his grades. You find that he has an 85% in math, an 80% in history, and a 75% in English. Why is it inappropriate to conclude that Jerry is doing much better in math than he is in English? Use the z score to frame your response.

SPSS QUESTIONS

1. Enter the following data into SPSS: 23, 33, 42, 47, 51, 61, 63, 67, 69, 71. Now, calculate the corresponding z scores. What are your results?

JUST FOR FUN/CHALLENGE YOURSELF

1. First, look up the equation for the normal curve. Now, using this equation, calculate the area under the curve for a z score between 0 and 1.

2. You measure the height of 10 people and come up with an average of 70 inches and a standard deviation of 4 inches. Calculate the t scores for the following cases: 68 inches, 76 inches.

3. If a variable has a mean of 53, a median of 62, and a standard deviation of 4, what is its skewness?

ANSWER KEY

TRUE/FALSE QUESTIONS

1. False. The percentages of scores under the normal curve are constant in the sense that they're the same regardless of the mean and standard deviation of the distribution.

2. True. Because z scores are standard scores, you can compare them across different distributions.

3. True. Values for the area under the normal curve can be represented as probabilities or percentages.

4. False. Because the normal curve is symmetrical, it doesn't matter whether the z score you're looking up is positive or negative. The area under the curve from the mean to a certain z score will be identical for the positive and negative version of that z score.

5. False. A standard score is very different from a standardized score. Standardized scores come from a distribution with a predefined mean and standard deviation, like the distribution of scores on the SAT or GRE.

6. False. This describes a midpoint.

7. True. $z = \dfrac{(X - \bar{X})}{s}$ $-2.33 = \dfrac{(3-10)}{3}$

8. False. $z = \dfrac{(X - \bar{X})}{s}$ $1.53 = \dfrac{(148 - 125)}{15}$ The area between the mean and a z score of 1.53 is 4.493.70. However, you must add 43.70 to 50 from the other side of the curve, so the correct answer is 93.7 percent.

9. False. Roger's z score is $-0.40 = \dfrac{(76 - 80)}{10}$. James' z score is $1.20 = \dfrac{(92 - 80)}{10}$. The area between the mean and a z score of .40 for Roger is 15.54 while the area between the mean and a z score of 1.20 for James is 38.49. Thus, 54.03 scores fall between the two means (38.49 + 15.54 = 54.03). Thus, 54.03 percent is correct.

10. False. A normal curve is asymptotic, which means that although it may come close, the curve never touches the horizontal axis.

MULTIPLE-CHOICE QUESTIONS

1. (c) 50%

2. (d) 100%

3. (c) 68.26%

4. (d) 95.44%

5. (b) 0

6. (b) positive

7. (b) 4

8. (c) 2.28%

9. (a) 30.85%

10. (d) 15.87%

11. (a) 1.68%

12. (a) 31.61%

13. (b) .05

14. (a) Skewness

15. (b) Kurtosis

16. (b) platykurtic

17. (b) .0107. The z score is 2.30, representing a percentile area of 48.93 between the score of 302 and the mean. 48.93 + 50 = 98.93, and 100 − 98.93 = 1.07 percent, or the probability of .0107.

18. (a) 33%

EXERCISES

1. The four z scores:

$$z = \frac{X - \bar{X}}{s} \Rightarrow \frac{2.1 - 5.8}{2.3} = -1.61$$

$$z = \frac{X - \bar{X}}{s} \Rightarrow \frac{5.7 - 5.8}{2.3} = -0.04$$

$$z = \frac{X - \bar{X}}{s} \Rightarrow \frac{7.3 - 5.8}{2.3} = 0.65$$

$$z = \frac{X - \bar{X}}{s} \Rightarrow \frac{12.4 - 5.8}{2.3} = 2.87$$

2. To achieve a score in the top 5%, you would first need to find the z score such that exactly 5% of scores lies between that z score and the highest possible score (or "infinity"). In other words, 95% of scores are below this score. Therefore, the z score that we are looking for is well above the mean. We know that 50% of scores lie below the mean, so you need to find a z score such that the area between the mean and the z score is 0.45 (this plus 0.50 equals 0.95 or 95%). This corresponds to a z score of approximately 1.645.

Now calculate:

$$X = z(s) + \bar{X} \Rightarrow 1.645(100) + 1,000 = 1,164.5$$

3. The five raw scores:

$$X = z(s) + \bar{X} \Rightarrow -5.3(4.2) + 27 = 4.74$$

$$X = z(s) + \bar{X} \Rightarrow -2.1(4.2) + 27 = 18.18$$

$$X = z(s) + \bar{X} \Rightarrow -0(4.2) + 27 = 27$$

$$X = z(s) + \bar{X} \Rightarrow 1(4.2) + 27 = 31.2$$

$$X = z(s) + \bar{X} \Rightarrow 3.1(4.2) + 27 = 40.32$$

SHORT-ANSWER/ESSAY QUESTIONS

1. In this chapter, the examples of IQ and height were presented. Exam scores, as well as final class scores, could also be examples of normally distributed measures. While the mean would probably be around a grade of C, you'd expect to see a smaller number of students get very high grades and a smaller number of students get very low ones. Health is another example. You'd expect most people to have typical health: pretty good, with perhaps some minor problems but no major ones. However, smaller percentages of individuals will have very poor health or very excellent health. In essence, the distribution of any normally distributed measure needs to look like the "bell curve": You have a large hump representing typical cases, with a peak equal to the mean. However, you also have smaller numbers of individuals who have more extreme scores, in both the positive and negative tails of the curve.

2. Comparing grades in two different classes fails to take into consideration the difficulty of the courses, the teaching styles of the instructors, and the population of the classes, among other factors. It could be that Jerry scores higher than most of the children in his English class, and is actually among the better students in that class. Perhaps all of the students in his math class are scoring around 85%, so he is neither better nor worse than most students. To better compare classes, you should determine Jerry's z score in each course, which are more comparable. If you find his z score in English is 1.4 and his z score in math is .80, this indicates that he is actually doing much better in English than in math.

SPSS QUESTIONS

1. The z scores are presented in the final column in the following screenshot:

	Var1	Zvar1
1	23.00	−1.81871
2	33.00	−1.20635
3	42.00	−.65522
4	47.00	−.34905
5	51.00	−.10410
6	61.00	.50826
7	63.00	.63073
8	67.00	.87567
9	69.00	.99815
10	71.00	1.12062

JUST FOR FUN/CHALLENGE YOURSELF

1.

$$\text{Area from } z_x \text{ to } z_y = \frac{1}{\sqrt{2\pi}} \int_x^y \Bigg|^{\frac{-z^2}{2}} dz$$

$$= \frac{1}{\sqrt{2\pi}} \int_0^1 \Bigg|^{\frac{-z^2}{2}} dz$$

$$= \frac{1}{\sqrt{2\pi}}(.8556)$$

$$= .3413$$

2. First, we calculate the z scores:

$$z = \frac{X - \bar{X}}{s} \Rightarrow \frac{68 - 70}{4} = -0.5$$

Then,

$$T = z \times 10 + 50 = -0.5 \times 10 + 50 = 45$$

$$T = z \times 10 + 50 = 1.5 \times 10 + 50 = 65$$

3.

$$Sk = \frac{3(\bar{X} - M)}{s} \Rightarrow \frac{3(53 - 62)}{4} = \frac{3(-9)}{4} = -6.75.$$

9 Significantly Significant

What It Means for You and Me

CHAPTER OUTLINE

LEARNING OBJECTIVES

- Understand the concept of statistical significance.

- Learn the difference between Type I errors and Type II errors.

- Understand the purpose of inferential statistics.

- Understand the distinction between statistical significance and meaningfulness.

- Learn the eight steps used to apply a statistical test to test any null hypothesis.

- Learn what confidence intervals are.

SUMMARY/KEY POINTS

- Statistical tests are based on probability: You are able to say with a certain level of certainty that there is a difference between groups or a relationship between variables, but you can't say this with 100% absolute certainty.
  - There is the possibility of making an error in judgment. This is why you should not say that you proved your hypothesis correct. It is possible to support or not support a hypothesis, but because chance or error might provide a better explanation for the outcome, we cannot say we proved a hypothesis.
  - The level of chance or risk that you are willing to take is expressed as a significance level.
  - A significance level of .05 corresponds to a 1 in 20 chance that any differences or relationships found based on statistical tests are not due to the hypothesized reason but are instead due to chance.
  - Researchers should try as much as possible to reduce this likelihood by removing all competing reasons for any differences or relationships. For example, if participants need to concentrate on their task, you should use a distraction-free room. If you think the gender or clothing of the experimenter might impact participants, then you should use the same experimenter for all sessions and have her dress the same way. However, error cannot be fully controlled because it is impossible to control for every possible factor.
  - There is always the possibility of error in statistics because the population itself is not directly tested. The sample is tested, and the results are inferred or generalized to the larger population. This inferential process always includes the possibility of error.

- A Type I error occurs when you reject the null hypothesis when there is actually no difference between groups or relationships between variables. For example, you might conclude that an authoritative parenting style leads to more confident, independent, and responsible children than a permissive parenting style. If there are no actual differences yet you conclude there are, you are making a Type I error.
  - The level of statistical significance is equal to the possibility of making a Type I error.
  - Type I errors are represented by the Greek letter alpha, or α.
  - These significance levels are typically set between .01 and .05, with .05 being the most common standard used. That is, you might conclude that authoritative parenting styles lead to more confident children than permissive parenting styles, but you recognize there is a 5% chance this conclusion is in error.
  - A statistical test that is nearly significant can be called "marginally significant." For example, if the probability level is set at .05 and the significance of your result is, say, .052 or .055, it can be reported as a "marginally significant" result.

- A Type II error occurs when you accept a false null hypothesis. Here, you might conclude that both permissive and authoritarian parenting styles lead to confident children, though in reality one parenting style may produce more confident children than the other.
  - This means that you conclude that there is no difference between groups or no relationship between variables when in fact there actually is.
  - Type II errors are represented by the Greek letter beta, or β.
  - Type II errors are not directly controlled for. However, they are related to factors such as sample size; Type II errors decrease as the sample size increases.

- Type I and Type II errors cover the scenarios in which errors are made; there are also two scenarios in which you make the correct judgment.
  - You can accept the null hypothesis when the null hypothesis is actually true. This means that you say there is no difference between groups, or no relationship between variables, and you are correct. That is, you correctly conclude that there are no differences between authoritarian and permissive parenting styles on the confidence of children when in fact there are no differences.
  - You can reject the null hypothesis when the null hypothesis is actually false. This means that you say there is a real difference between groups, or relationship between variables, and you are correct. That is, you correctly find that authoritarian parenting styles lead to more confident children than permissive parenting styles when this is, in fact, true.
  - This is also called power, or $1 - \beta$. In other words, power is equal to the value of the Type II error subtracted from 1.

- There is an important difference between statistical significance and meaningfulness. It is possible to have a result that is statistically significant but so small that it is not really meaningful. Determining meaningfulness depends in part on the context of the issue and design of the study. Even though you might find higher confidence among children of authoritative parents compared to permissive parents, this higher confidence does not imply a better chance at success. Higher confidence might even have negative consequences, as children are unable or unwilling to alter their perspectives, even when they are incorrect.

- While descriptive statistics describe data (by means of tables, charts, etc.), inferential statistics are used to infer something about the population based on a sample's characteristics.
  - The practice of inferential statistics uses a wide variety of statistical tests and analyses to test differences between groups or relationships between variables. For example, we could run a t-test to determine if a child's confidence differs depending on whether an authoritarian or permissive parent raised her.
  - Tests of significance are used in inferential statistics. These tests of significance are based on the fact that each null hypothesis can be tested with a particular type of statistical test. Every calculated statistic has a special distribution associated with it. The calculated value is then compared to the distribution to conclude whether the sample characteristics are different from what you would expect by chance.

- Applying a statistical test to any null hypothesis follows eight general steps:
  1. Provide a statement of the null hypothesis.
  2. Set the level of risk associated with the null hypothesis (significance level).
  3. Select the appropriate test statistic.
  4. Compute the test statistic value (also known as the obtained value).
  5. Determine the value (the critical value) needed for rejection of the null hypothesis using the appropriate table of critical values for that particular statistic.

6. Compare the obtained value with the critical value.

7. If the obtained value is more extreme than the critical value, the null hypothesis must be rejected.

8. If the obtained value does not exceed the critical value, the null hypothesis cannot be rejected.

- Confidence intervals represent the best estimate of the range of the population value (or population parameter) based on the sample value (or sample statistic).
  - A higher confidence interval (for example, a 99% confidence interval as compared with a 95% confidence interval) represents a greater degree of confidence, meaning that a wider range of values will be incorporated into the confidence interval.

KEY TERMS

- **Significance level** or **statistical significance**: The level of risk set by the researcher for rejecting a null hypothesis when it is true. In other words, it corresponds to the level of risk that there is actually no relationship between variables when the results of your analysis appear to tell you that there is.

- **Type I error**: The rejection, or the probability of rejection, of a null hypothesis when it is true

- **Type II error**: The acceptance, or the probability of acceptance, of a null hypothesis when it is false

- **Inferential statistics**: A set of tools that are used to infer the results of an analysis based on a sample to the population

- **Test statistic** (aka obtained value): The value that results from the use of a statistical test

- **Critical value**: The value necessary for rejection (or nonacceptance) of the null hypothesis

- **Confidence interval**: The best estimate of the range of a population value given the sample value

TRUE/FALSE QUESTIONS

1. A 99% confidence interval has a larger range than a 95% confidence interval for the same analysis.

2. Using inferential statistics, it is common to say that you have 100% confidence in a result.

3. A psychologist concludes that women are better able to repress memories of childhood trauma than men. She has made a Type II error in her conclusion if the data actually shows no differences among men and women.

4. It is much more important to have statistical significance than meaningfulness.

5. It is possible to control all potential error in a research study.

MULTIPLE-CHOICE QUESTIONS

1. A significance level of .05 corresponds to a _____ chance that any differences or relationships found based on a statistical test are not due to the hypothesized reason but are instead due to chance.

   a. 1 in 10

   b. 5 in 10

   c. 1 in 5

   d. 1 in 20

2. Which of the following relates to the rejection of the null hypothesis when it is actually false?

   a. Type I error

   b. Type II error

   c. Power

   d. Statistical significance

3. Which of the following relates to the acceptance of the null hypothesis when it is actually true?

   a. Type I error

   b. Type II error

   c. Power

   d. A correct decision

4. Which of the following relates to the rejection of the null hypothesis when it is actually true?

   a. Type I error

   b. Type II error

   c. Power

   d. Statistical significance

5. Which of the following relates to the acceptance of the null hypothesis when it is actually false?

   a. Type I error

   b. Type II error

   c. Power

   d. Statistical significance

6. Which of the following is calculated by subtracting the Type II error from 1?

   a. Type I error

   b. Type II error

   c. Power

   d. The significance level

7. Type I errors are represented as _____.

    a. α

    b. β

    c. 1 − β

    d. *e*

8. Type II errors are represented as _____.

    a. α

    b. β

    c. 1 − β

    d. *E*

9. You conduct a study to see whether there is a significant difference in IQ between two classes. The first class has an average IQ of 114.1, while the second class has an average IQ of 114.7. You conduct a statistical test, which finds a significant difference between these two groups at the .05 level of significance. In sum, these results are _____.

    a. statistically significant only

    b. meaningful only

    c. statistically significant and meaningful

    d. neither statistically significant nor meaningful

10. When conducting a statistical test, you set the significance level at .05. After running the analysis, you find a significance level of .054. This result is _____.

    a. statistically significant

    b. marginally significant

    c. meaningful only

    d. none of the above

11. In inferential statistics, you infer from a _____ to a _____.

    a. larger population; smaller sample

    b. smaller sample; larger population

    c. smaller population; larger sample

    d. larger sample; smaller population

12. In terms of *z* scores, a 95% confidence interval consists of which of the following ranges?

    a. ±1 *z*

    b. ±1.96 *z*

    c. ±2.05 *z*

    d. ±2.56 *z*

13. In terms of z scores, a 99% confidence interval consists of which of the following ranges?

    a. $\pm 1\ z$

    b. $\pm 1.96\ z$

    c. $\pm 2.05\ z$

    d. $\pm 2.56\ z$

14. If a class took an exam and scored a mean of 82 with a standard deviation of 12, what would be the 95% confidence interval?

    a. $12 \pm 2.56(82)$

    b. $82 \pm 2.56(12)$

    c. $12 \pm 1.96(82)$

    d. $82 \pm 1.96(12)$

15. Using the .05 level of significance, which of the following findings is statistically significant?

    a. Higher levels of crime have been found in cities with greater population density ($p < .05$).

    b. Global temperatures have been found to be increasing steadily since 1900 ($p = .053$).

    c. Crime rates in the United States have been steadily decreasing since 1990 ($p = .06$).

    d. No gender differences in scores were found ($p = .12$).

16. Which of the following significance levels gives you the greatest chance of finding a significant result?

    a. .10

    b. .05

    c. .01

    d. .001

17. Which of the following values results from the use of a statistical test?

    a. The critical value

    b. The obtained value

    c. Type I error

    d. Type II error

18. Which of the following is the value necessary for rejection (or nonacceptance) of the null hypothesis?

    a. The critical value

    b. The obtained value

    c. Type I error

    d. Type II error

19. Which of the following is the best estimate of the range of a population value given the sample value?

    a. The critical value

    b. The obtained value

    c. Power

    d. Confidence interval

20. Contrary to predictions, a psychologist finds that positive punishment (applying something aversive following unwanted behavior) does not differ from negative punishment (taking away something desirable). Retaining the null hypothesis here assumes which of the following:

    a. There is no difference between sample means.

    b. There is no difference between population means.

    c. The difference between means is significant.

    d. The difference between the population mean and the sample mean is significant.

21. A psychologist makes a Type II error when predicting that children who witness an adult swearing will swear more often themselves than children who witness an adult talking conversationally. What does this mean?

    a. The psychologist says there was a difference when there was no difference.

    b. The psychologist says there was a difference when there was a difference.

    c. The psychologist says there was no difference when there was a difference.

    d. The psychologist says there was no difference when there was no difference.

22. A researcher predicts that participants will score higher on an exam if they work in pairs rather than when they work in threesomes. After noting that the null hypothesis entails no differences between the two conditions, setting her level of risk at .05, and selecting the appropriate test statistic (in this case a *t*-test), what should she do next?

    a. Compute the test statistic value.

    b. Determine the value needed to reject the null hypothesis using the critical value for that particular statistic.

    c. Compare the obtained value with the critical value.

    d. Decide if the obtained value exceeds the critical value.

EXERCISE

1. Come up with a hypothesis in an area of study that interests you. Now, conduct a "mock" statistical test, writing out the eight steps you would use to apply a statistical test to your hypothesis.

SHORT-ANSWER/ESSAY QUESTION

1. What's an example of a result that is statistically significant but not meaningful? Why is it not meaningful?

2. Why shouldn't a researcher set the confidence level at .0001, since it is much smaller than .05 and even smaller than .01? Answer this question by taking into consideration Type I and Type II errors.

3. Imagine you predict that it will take a salesclerk at a mall longer to help out a potential customer who is dressed in sloppy clothes (untucked t-shirt and torn jeans) than a potential customer dressed in business attire. You run the study and find support for your hypotheses, with the obtained value exceeding the critical value at the $p < .05$ level. Why wouldn't you want to say that you proved your hypothesis correct? What other sources might be a better explanation for your findings?

JUST FOR FUN/CHALLENGE YOURSELF

1. Spend a few minutes reading about statistical power online or in a statistics book. Next, download G*Power, a free software program for power calculations, or find an online calculator that can calculate the power for correlation coefficients. Using a Pearson correlation (bivariate normal model), what is the sample size needed for a two-tailed test if the null correlation (r) is 0, the research hypothesis correlation (r) is 0.5, your level of significance is .05, and you want a power of 0.9?

ANSWER KEY

TRUE/FALSE QUESTIONS

1. True. To have a higher level of confidence (i.e., 99% confidence instead of 95% confidence), you would need to incorporate a larger set of values into the range of the confidence interval. This means that a 99% confidence interval must have a larger range than does a 95% confidence interval.

2. False. Inferential statistics uses probability, which in this case means that you can never be completely certain of a result.

3. False. She says the two groups differed when they did not differ, a Type I error.

4. False. You might find significance in a study, but it might not be meaningful. For example, a costly workplace training program might improve work productivity, but the benefits of the productivity may not exceed the costs of the training program.

5. False. There will always be some degree of error in a study. There might be experimenter error, error based on environmental factors (like distractions or uncomfortable temperatures), or measurement errors, among many others. We do not need to be 100% confident in research. Rather, we give ourselves some leeway in our confidence by permitting 5% error (or less) but still concluding that results are statistically significant.

MULTIPLE-CHOICE QUESTIONS

1. (d) 1 in 20

2. (c) Power

3. (d) A correct decision

4. (a) Type I error

5. (b) Type II error

6. (c) Power

7. (a) α

8. (b) β

9. (a) statistically significant only

10. (b) marginally significant

11. (b) smaller sample; larger population

12. (b) $\pm 1.96\ z$

13. (d) $\pm 2.56\ z$

14. (d) $82 \pm 1.96(12)$

15. (a) Higher levels of crime have been found in cities with greater population density ($p < .05$).

16. (a) .10

17. (b) The obtained value

18. (a) The critical value

19. (d) Confidence interval

20. (b) There is no difference between population means.

21. (c) The psychologist says there was no difference when there was a difference.

22. (a) Compute the test statistic value.

EXERCISES

1. As an example, let's say that I'm interested in crime rates and want to study whether there is a significant difference in crime rates (per capita) between the United States and Europe. Your answer does not have to be as detailed as mine, and we will get more detailed and specific in the next few chapters. Here are the eight steps that we would use to test this:

   1. A statement of the null and research hypotheses:

      The null hypothesis: $H_0: \mu_1 = \mu_2$.

      The research hypothesis: $H_1: \bar{X}_1 \neq \bar{X}_2$.

   2. Set the level of risk, or significance, associated with the null hypothesis: 0.05.

   3. Select the appropriate test statistic. In this case, we would use the t-test for independent means, because we are testing the difference between two separate groups.

   4. Compute the test statistic value (obtained value). This goes beyond what we've covered so far, but we will fully complete this step in later chapters, which focus on particular statistical tests. For the sake of argument, say that our obtained value is 3.7.

5. Determine the critical value. Again, to continue with this example, say that our critical value is 2.021.

6. Now, compare the obtained value with the critical value. As you can see, our obtained value is larger than our critical value.

7. Because the obtained value is more extreme than the critical value, we can say that our null hypothesis cannot be accepted.

8. Because the obtained value is larger than the critical value, we must reject the null hypothesis.

SHORT-ANSWER/ESSAY QUESTION

1. A result can be statistically significant but not meaningful. Imagine your p value is set at less than .05, and you compare grades for two psychology exams, where the class average on the first exam was 87.2 and the class average on the second exam was 88.1. Although these average scores might be statistically different at $p < .05$, the difference between 87.2 and 88.1 may not be very meaningful. You might also have a significant but not meaningful relationship if the correlation between two variables had a significance level below .05 but the strength of the correlation was very weak.

    The reason why these examples do not illustrate a meaningful is that the group difference or strength of the relationship is extremely low.

2. A researcher should not set the risk level at .0001 since it is too small. It suggests that there is a 1 in 10,000 chance (a REALLY small risk) that you will reject the null hypothesis (there are no differences) when it is actually true. This makes it very likely that a Type II error will occur, as you might incorrectly conclude that the groups in the study did not differ when they actually did differ, and would have at a lower p value.

3. We can never be 100% sure our independent variable alone is responsible for scores on the dependent variable, as there will always be some degree of error. Researchers allow a little bit of error in the results, noting that as long as that error is less than 5%, the hypothesis is still supported. In the clothing study, it is possible that salesclerks in the sloppy clothes condition were distracted by other customers. If this is the case, they may simply have not seen the sloppy customer as quickly as nondistracted salesclerks in the business attire condition, which is a better explanation for differences between conditions than customer attire.

JUST FOR FUN/CHALLENGE YOURSELF

1. The minimum sample size needed in this example is 37.

10

Only the Lonely

The One-Sample z-Test

LEARNING OBJECTIVES

- Understand when it is appropriate to use the one-sample z-test.

- Learn how to compute the observed z value for a one-sample z-test.

- Learn how to interpret the z value and understand what it means.

- Be able to work through the eight steps of testing a hypothesis when using the one-sample z-test.

- Learn the basic concepts of effect size.

SUMMARY/KEY POINTS

- A one-sample z-test is used to compare the mean of a sample to the mean of a population.
  - This test is used when only one group is being tested. For example, a researcher might be interested in looking at Post-Traumatic Stress Disorder (PTSD) rates for soldiers returning from the battlefield in Iraq and Afghanistan in comparison to the population of soldiers. If the incidents of PTSD among those in the battlefield is around 12.5% and the incidents for the population of soldiers is 5.5%, the question becomes, "Is the battlefield sample representative of the population of soldiers?"
  - The obtained value is determined by calculating the one-sample z-test.
  - After the z value is calculated, the critical value is obtained from a table of z scores so that a comparison can be made.
  - Effect size can help you to understand whether a statistically significant result is also meaningful.
    - A small effect size ranges from 0 to .2.
    - A medium effect size ranges from .2 to .5.
    - A large effect size is any value above .5.

KEY TERMS

- **One-sample z-test**: A statistical test used to compare a sample mean to a population mean

- **Standard error of the mean**: An error term that is used as the denominator in the equation for the z value in a one-sample z-test. The standard error of the mean is the standard deviation of all possible means selected from the population.

- **Effect size**: A measure of how different two groups are from one another; it is a measure of the magnitude of a treatment

TRUE/FALSE QUESTIONS

1. A one-sample z-test can be used to compare the mean of two populations.

2. For a one-sample z-test to be significant at the .05 level of significance, you need an obtained z value of at least 1.96 (or below -1.96).

3. A sports psychologist wants to see if the number of concussions sustained by the Denver Broncos is representative of the population of concussions in the National Football League. The one-sample z-test is appropriate for this analysis.

4. To assess the impact of pain on memory, a researcher has some participants place their hand in ice-cold water for ten seconds while others place their hand in 104-degree water (the maximum temperature recommended for hot tubs). It is appropriate for the researcher to run a one-sample z-test on the resulting memory scores.

MULTIPLE-CHOICE QUESTIONS

1. A one-sample z-test would be used in which one of the following situations?

   a. Comparing two sample means

   b. Comparing two population means

   c. Comparing a sample mean with a population mean

   d. Comparing two sample means with a population mean

2. In a one-sample z-test, _____ groups are being tested.

   a. one

   b. two

   c. three

   d. two or more

3. In the equation for a one-sample z-test, the denominator is known as _____.

   a. the sample mean

   b. the population mean

   c. the standard error of the mean

   d. the standard error of the population

4. Which of the following values represents the standard deviation of all the possible means selected from the population?

   a. The sample mean

   b. The population mean

   c. The standard error of the mean

   d. The standard error of the population

5. You conducted a one-sample z-test, obtaining a value of 4.6 for the z value. What is your conclusion?

   a. I should reject the null hypothesis.

   b. I should accept the null hypothesis.

   c. The result is too close to call.

   d. None of the above is correct.

6. A researcher is concerned that women may feel working in Information Technology (IT) will leave them feeling isolated, making it harder to attract women into this growing field. She knows that, on average, the population of IT workers rate their isolation at 2.5 on a 0-to-10 scale, with higher scores indicating more isolation. The standard deviation is 1.20. She asks a sample of 25 women working in IT to rate their feelings of isolation, and finds that they rate it an average of 3 with a standard deviation of 0.80. Should the researcher conclude that the sample of women differs from the general population of IT workers?

   a. Yes, the sample and the population differ.

   b. No, the sample and the population do not differ.

   c. The result is too close to call.

   d. None of the above is correct.

7. A researcher looks at panic attacks among a sample of women at a small community college. In comparison to the population of college students enrolled at the college, she finds $z = 1.23, p > .05$. What should the researcher conclude?

   a. The sample is representative of the college population.

   b. The sample is not representative of the college population.

   c. The sample has more panic attacks than the population.

   d. The sample has less panic attacks than the population.

8. A researcher finds a z score of 1.99, $p < .05$ for a sample of participants rating their own emotional intelligence compared to the emotional intelligence scores of the population, with an effect size of 0.023. What does this mean?

   a. The z-test is both significant and meaningful.

   b. The z-test is not significant but meaningful.

   c. The z-test is significant but not meaningful.

   d. The z-test is neither significant nor meaningful.

EXERCISES

1. You are a teacher at a gifted school, and you feel that the newest class of students is even brighter than usual. The mean IQ at your school is 127, and the mean IQ of this new class is 134. In total, there are 32 students in this new class. Also, the standard deviation of the school's IQ is 8. Use the eight steps to test whether this new class of students is significantly more intelligent than the school's student body overall.

2. Interpret the following result: $z = 3.49, p < .05$.

SHORT-ANSWER/ESSAY QUESTION

1. Come up with two examples in which a one-sample z-test would be appropriate.

2. State the null and research hypotheses for the examples you came up with in question #1 above.

3. Imagine a researcher finds that a sample of children with behavioral and mental problems is significantly higher in one neighborhood than the rest of the city, $z = 2.33$, $p < .05$. What factors might account for the high rate of disorders in that one neighborhood compared to the rest of the city?

JUST FOR FUN/CHALLENGE YOURSELF

1. A one-sample z-test was conducted in which the sample mean was 76, the population mean was 82, the population standard deviation was 2, and the calculated z value was -15. What was the sample size?

ANSWER KEY

TRUE/FALSE QUESTIONS

1. False. The one-sample z-test is used to compare the mean of a sample with the mean of a population, not to compare the mean of two populations.

2. True. At the .05 level of significance, a two-tailed test has a critical value of 1.96 for the z value.

3. True. The researcher is comparing the sample values (concussions among Denver Bronco players) with the population (concussions among all NFL players). A one-sample z-test is appropriate here.

4. False. The study describes a two-group design rather than comparing a sample to a population. The appropriate test would most likely be a t-test.

MULTIPLE-CHOICE QUESTIONS

1. (c) Comparing a sample mean with a population mean

2. (a) one

3. (c) the standard error of the mean

4. (c) The standard error of the mean

5. (a) I should reject the null hypothesis.

6. (a) Yes, the sample and the population differ. $Z = \dfrac{\bar{x} - \mu}{\dfrac{\sigma}{\sqrt{n}}} = \dfrac{3 - 2.5}{\dfrac{1.2}{\sqrt{25}}} = 2.08$. 2.08 is above the 1.96 threshold needed to show that the sample differs from the population, $z = 2.08$, $p < .05$.

7. (a) The sample is representative of the college population.

8. (c) The z-test is significant but not meaningful.

EXERCISES

1. Here are the eight steps used to test this hypothesis:

    1. The null hypothesis: H_0: $\bar{X} = \mu$

    2. The alternative hypothesis: H_1: $\bar{X} = \mu$

    3. The level of significance: 0.05.

    4. The appropriate test statistic: The one-sample z-test.

    5. Computation of the test statistic value:

    $$SEM = \frac{\sigma}{\sqrt{n}} \Rightarrow \frac{8}{\sqrt{32}} = 1.41$$

    $$z = \frac{\bar{X} - \mu}{SEM} \Rightarrow \frac{134 - 127}{1.41} = 4.96$$

    As this is a one-tailed test, the value needed for rejection of the null hypothesis is ±1.66. This represents the point at which only 5% of scores are higher than this value, corresponding to our significance level of .05. If we had been conducting a two-tailed test, the critical value would have been ±1.96.

    6. Comparing the obtained value with the critical value, we see that the obtained value is higher than the critical value.

    7. and 8. Because our obtained value is greater than the critical value, we can reject the null hypothesis, which suggests no difference between the sample and population mean. Our results show that this new class is significantly brighter than the school's students overall.

2. First, the z represents the test statistic that was used. The value of 3.49 represents the obtained z value that was calculated as part of conducting the one-sample z-test. Finally, $p < .05$ indicates that we have at least a 95% level of certainty that these two groups (the sample and the population) do differ in regard to their mean values.

SHORT-ANSWER/ESSAY QUESTION

1. These two examples could be any situation in which you are comparing a sample mean to a population mean. For example, a one-sample z-test would be appropriate if you were comparing reading-test scores in one school with scores of the entire nation. Another example could be testing whether the crime rate in a certain state or neighborhood is significantly different from the national crime rate.

2. For the reading-score example, the null hypothesis would be that the school and the nation do not differ. The research hypothesis would be that the school and the nation scores do differ. The same thing applies to the crime rate example. The null would state no difference in rates between the state or neighborhood and the national crime rate while the research hypothesis would predict a difference.

3. There might be both social and environmental factors that lead one neighborhood to have high rates of mental and behavioral disorders. For example, environmental factors like lead poisoning may lead to problems among children, including central auditory processing problems, trouble finding words, lower IQs, and slower response times. Yet social factors might also account for higher rates. Parents might want to connect with other parents who have children with similar behavior and mental disorders, or there might be specialized services available in a specific neighborhood that draws parents who need aid to that community.

JUST FOR FUN/CHALLENGE YOURSELF

1. We can solve this using the equation for the one-sample z-test:

$$z = \frac{\bar{X} - \mu}{\dfrac{\sigma}{\sqrt{n}}} \Rightarrow$$

$$-15 = \frac{76 - 82}{\dfrac{2}{\sqrt{n}}}$$

$$-15 = \frac{-6}{\dfrac{2}{\sqrt{n}}}$$

$$-15 = -6\left(\frac{\sqrt{n}}{2}\right)$$

$$-15 = -3\sqrt{n}$$

$$5 = \sqrt{n}$$

$$n = 25$$

11 *t*(EA) for Two

Tests Between the Means of Different Groups

LEARNING OBJECTIVES

- Understand when it is appropriate to use the *t*-test for independent means.

- Learn how to calculate the observed *t* value by hand and with SPSS.

- Learn how to interpret the results of a *t*-test.

- Understand the difference between a significant and meaningful result and how this relates to the effect size.

SUMMARY/KEY POINTS

- Use the *t*-test for independent means when you are looking at the difference between the mean scores of two groups on one or more variables and the two groups are independent of one another (i.e., are not related in any way).
  - This test is used when each group is tested only once and the dependent variable is normally distributed and based on an interval or ratio scale.
  - There must be only two groups in total.
  - The corresponding test statistic is the *t*-test for independent means.
  - For example, a researcher might be interested in the effects of self-focus on partici- pants' likelihood of using personal pronouns (I, me, myself). He stops pedestrians and asks them if he can interview them. For some participants, he takes their picture. For others, he proceeds directly to the interview. After coding the interview responses, he can use a *t*-test to determine if those in the picture condition used more personal pro- nouns than those in the no-picture condition.

- After the *t*-test for independent means is conducted, compare the obtained value with the critical value to see whether you have statistical significance. Consider the self-focus study. If the *t*-test obtained value exceeds the critical value, then the researcher can con- clude that the picture and no-picture groups differed. Of course, the researcher will then need to look at the mean responses on the dependent variable (the average number of personal pronouns used) to see if participants in the picture condition used more or fewer personal pronouns than those in the no-picture condition.

- There's an important difference between a significant result and a meaningful result. A component of a result's meaningfulness is effect size, which is a measure of the magni- tude of the treatment.

- When *t*-tests are used, the measure of effect size is Cohen's *d*.
  - A small effect size ranges from 0 to .20.
  - A medium effect size ranges from .20 to .50.
  - A large effect size is any value above .50.
  - A larger effect size represents a greater difference between the two groups.

KEY TERMS

- **Homogeneity of variance assumption**: An assumption underlying the *t*-test that the amount of variability in both groups is equal

- **Degrees of freedom**: A value that approximates the sample size and is a component of many statistical tests

- **Effect size**: A measure of how different groups are from one another (i.e., a measure of the magnitude of the effect)

- **Pooled standard deviation**: Part of the formula for the effect size that is similar to an average of the standard deviations from both groups

TRUE/FALSE QUESTIONS

1. The observed value $t_{(24)} = 2.35$ is significant at the .05 level (two-tailed test).

2. The t-test for independent samples should be used when participants are tested multiple times.

3. The sign of the observed t value (i.e., whether it is positive or negative) is always a crucial element in conducting the t-test for independent samples.

4. A researcher finds that participants who frequently post on Facebook rate their feelings of loneliness higher than those who post infrequently on Facebook. Since the researcher did not randomly assign participants to frequent versus infrequent Facebook posting, he cannot analyze loneliness using an independent samples t-test.

5. A psychologist tests whether hunger impairs memory performance by having some participants fast for two days prior to a memory study and other participants eat three balanced meals for each of the two days prior to a memory study. He finds $t(35) = 2.13$, $p < .05$, and thus correctly concludes that those who fasted ($M = 2.34$, $SD = 1.13$) performed significantly worse on the memory test than those who did not fast ($M = 4.55$, $SD = 1.43$).

6. Researchers can use a t-test to evaluate nominal variables, like whether a potential patient has or does not have a mood disorder.

7. A researcher studying color blindness finds that a color-blind participant correctly chooses the correct color from a color palette more often when wearing a special pair of glasses than when not wearing glasses. The correct test to run in this study is an independent samples t-test.

MULTIPLE-CHOICE QUESTIONS

1. Which of the following is significant at the .05 level (two-tailed test)?

    a. $t_{(32)} = 2.01$.

    b. $t_{(212)} = 1.99$.

    c. $t_{(6)} = 2.30$.

    d. $t_{(2)} = 4.10$.

2. Which of the following is significant at the .01 level (one-tailed test)?

    a. $t_{(1)} = 28.90$.

    b. $t_{(16)} = 2.42$.

    c. $t_{(70)} = 2.45$.

    d. $t_{(80)} = 2.37$.

3. The t-test for independent samples should be used in which of the following scenarios?

    a. You are comparing more than two groups that are related.

    b. You are comparing exactly two groups that are related.

    c. You are comparing exactly two groups that are unrelated.

    d. You are comparing more than two groups that are unrelated.

4. The assumption that the *t*-test for independent samples makes regarding the amount of variability in each of the two groups is called the _____.

   a. homogeneity of variance assumption

   b. equality of variance assumption

   c. assumption of variance equivalence

   d. variability equality assumption

5. The value used in the equation for the *t*-test for independent means that approximates the sample size is known as the _____.

   a. homogeneity of variances

   b. sample size equivalence

   c. degrees of freedom

   d. standard deviation

6. Imagine that you look up a critical *t* value in a *t* table and your observed *t* value for a specific degrees of freedom lies between two critical values in the table. To be conservative, you would select the _____.

   a. smaller value

   b. larger value

   c. smallest value in the table

   d. largest value in the table

7. After conducting a *t*-test for independent samples, you arrived at the following: $t_{(47)} = 3.41, p < .05$. The degrees of freedom is indicated by which of the following?

   a. 47

   b. 3.41

   c. < .05

   d. *p*

   e. *t*

8. After conducting a *t*-test for independent samples, you arrived at the following: $t_{(47)} = 3.41, p < .05$. The *t* value is indicated by which of the following?

   a. 47

   b. 3.41

   c. < .05

   d. *p*

   e. *t*

9. After conducting a *t*-test for independent samples, you arrived at the following: $t_{(47)} = 3.41, p < .05$. The probability is indicated by which of the following?

   a. 47

   b. 3.41

   c. < .05

   d. *t*

10. After conducting a *t*-test for independent samples, you arrived at the following: $t_{(47)} = 3.41$, $p < .05$. The test statistic that was used is indicated by which of the following?

    a. 47

    b. 3.41

    c. $< .05$

    d. *p*

    e. *t*

11. An effect size of .25 would be considered _____.

    a. small

    b. medium

    c. large

12. An effect size of .17 would be considered _____.

    a. small

    b. medium

    c. large

13. An effect size of .55 would be considered _____.

    a. small

    b. medium

    c. large

14. If there is no difference between the distributions of scores in two groups, your effect size will be equal to which of the following?

    a. 0

    b. 0.1

    c. 0.5

    d. 1

15. Temperature has a direct correlation with aggression. Knowing this, a researcher predicts that participants completing a frustrating task will express more aggression when they are in a 90-degree room than a 70-degree room. The researcher finds $t_{(20)} = 2.33, p < .05$. What can the researcher conclude?

    a. Participants were more aggressive in the 90-degree room than the 70-degree room.

    b. Participants were more aggressive in the 70-degree room than the 90-degree room.

    c. Participants were equally aggressive in the 70-degree room and 90-degree room.

    d. There is not enough information to render a full conclusion.

16. In a study with 40 participants, a researcher randomly assigns 20 participants to watch a brief film in which the protagonist helps someone in need. The remaining 20 participants watch a brief film in which the protagonist ignores someone in need. After the experiment ends, they see someone (actually a confederate) drop some papers, and the researcher watches to see how long it takes participants to help pick up the papers. Those who watch the "helping" film are quicker to provide assistance than those who saw the "ignoring" film. In this study, what is the degrees of freedom?

    a. 40

    b. 38

    c. 20

    d. There is not enough information to assess the degrees of freedom.

17. A researcher finds that participants rate their intention to donate to charity higher when the request comes from someone dressed as a priest than when the same person is dressed as a police officer. There are 25 people in each condition, and the obtained *t* value (two-tailed) is 2.85. What is the correct write up in this study?

    a. $t_{(50)} = 2.85, p < .05$

    b. $t_{(48)} = 2.85, p < .01$

    c. $t_{(48)} = 2.85, p < .05$

    d. $t_{(48)} = 2.85, p > .05$

18. A researcher wants to buy a new one-million-dollar MRI machine to replace the older model he bought five years earlier. Research shows that the new machine is significantly faster than the old machine $t_{(60)} = 2.10, p < .05$, with an effect size of .14. Should he buy the new costly machine?

    a. Yes. The new machine is significantly faster than the old machine and the effect size is large.

    b. Yes. The new machine is significantly faster than the old machine and the effect size is medium.

    c. Yes. The new machine is significant faster than the old machine and the effect size is small.

    d. No. While the new machine is significantly faster than the old machine, the effect size is too small to be meaningful.

EXERCISES

1. Using the door-in-the-face technique (a psychological technique that works because participants are more likely to comply with a small request after first rejecting a large request), a researcher asks participants in the door-in-the-face condition to volunteer two weeks of their time to help out at a summer camp for impoverished children. When, as expected, they decline this large request, he asks them if they would be willing to donate money to the camp instead (the small request). For participants in the control condition, the researcher asks them to donate money (the small request) without mentioning the two-week volunteer time. This is your data (in dollars donated for the camp):

Foot-in-the-door condition	Control condition
87	86
98	73
75	79
88	56
76	72
85	70
92	87
56	59
89	64
85	77
84	74
92	72

Use the eight steps to test the null hypothesis that there is no difference between the two conditions. Do the two groups differ?

2. Assuming that the standard deviations of two groups are equal and the standard deviation is 4.3, calculate the effect size if the two groups have means of 47.7 and 58.2. Is this effect size small, medium, or large?

3. Using the online calculator at www.uccs.edu/lbecker/, calculate the effect size using the data from the previous question. Do your results match?

4. Interpret the following result: $t_{(21)} = 28.90$, $p < .05$.

SHORT-ANSWER/ESSAY QUESTIONS

1. Imagine you are reading journal articles for a paper you are writing. The authors of Article One compared two groups on an outcome measure and found a statistically significant difference between the groups. The sample sizes of the groups were small. The authors of Article Two compared two other groups on a similar outcome measure. The results in this article were also statistically significant. However, the sample sizes in this study were not similar to each other. One group was quite large, and the other group was smaller but still large in comparison to the groups discussed in Article One. The p values in the articles were similar, but with such a discrepancy in sample size between the two articles, you are wondering whether the results from both studies are practically meaningful and not just statistically significant. What other result will you look for (or calculate) to answer your question? Explain why you chose that test result and discuss its features.

2. Please refer to page 218 of *Statistics for People . . .* and read the feature titled "So How Do I Interpret . . . ?" You will see that t is a negative number. Is the result statistically insignificant because t is negative? Please explain your answer. Then describe any other relevant information that could be provided to describe the data used in the analysis.

3. A researcher predicts that the color red will have a negative impact on participants' ability to solve a series of math problems. He gives twenty participants twenty math problems written in red ink while twenty other participants complete the same math problems written in black ink. The mean time for participants in the red ink condition to complete all math questions is 320 seconds (standard deviation is 63 seconds) while black ink participants complete the math task in an average of 282 seconds (standard deviation is 59 seconds), $t_{(38)} = 2.54$, $p < .05$. How would you interpret this study in terms of both statistical significance and meaningfulness? Use the effect size calculator available at http://www.uccs.edu/~lbecker/

4. A researcher is interested in the impact of three dosage levels of a new therapy designed to treat anxiety and depression. One treatment involves cognitive-behavioral therapy alone, one treatment involves antidepressants alone, and one treatment involves a combination of both cognitive behavioral therapy and antidepressants. Would it be appropriate to analyze the data from this study with an independent samples *t*-test. Why or why not?

SPSS QUESTION

1. Input the data under the "Exercises" section, question 1, into SPSS. Now, use SPSS to conduct a *t*-test for independent samples. How would you interpret the results? Do your results match those calculated by hand?

JUST FOR FUN/CHALLENGE YOURSELF

1. You study two apple orchards to see whether one produces significantly more fruit than the other. The first orchard produces an average of 14.2 tons of apples per year, while the second produces 17.3 tons per year, on average. The standard deviation of the amount of apples produced per year is 4.2 tons for the first orchard and 2.3 tons for the second orchard. Calculate the effect size using the equation that utilizes the pooled standard deviation.

ANSWER KEY

TRUE/FALSE QUESTIONS

1. True.

2. False. This test should only be used when participants are tested only once.

3. False. The sign of the observed *t* value is not important when the *t*-test for independent samples is nondirectional. The sign of the observed *t* value is important when the *t*-test for independent samples is directional.

4. False. The researcher may not be able to draw causal conclusions from the data, but he can still show that Group 1 (frequent Facebook posters) differ from Group 2 (infrequent Facebook posters) using an independent samples *t*-test.

5. True. The obtained value of 2.13 exceeds the critical value of 2.03 with degrees of freedom of 35 (two-tailed), and thus the groups differ. Since the mean score for the group that fasted ($M = 2.34$) is lower than the mean for the group that did not fast ($M = 4.55$), the conclusion that those who fasted performed more poorly is correct.

6. False. The t-test compares means. Since nominal variables like whether someone has or does not have a mood disorder are categorical in nature, they are not based on means, and therefore should not be analyzed with a t-test.

7. False. Because the researcher used the same participant more than once, the two groups are not independent. The more appropriate test is a dependent samples t-test, which is the focus of Chapter 12.

MULTIPLE-CHOICE QUESTIONS

1. (b) $t_{(212)} = 1.99$.

2. (c) $t_{(70)} = 2.45$.

3. (c) You are comparing exactly two groups that are unrelated.

4. (a) homogeneity of variance assumption

5. (c) degrees of freedom

6. (a) smaller value

7. (a) 47

8. (b) 3.41

9. (c) $< .05$

10. (e) t

11. (b) medium

12. (a) small

13. (c) large

14. (a) 0

15. (d) There is not enough information to render a full conclusion. Although the groups do differ significantly, without the means it is impossible to determine which group (70 degree or 90 degree) expresses more aggression.

16. (b) 38. Degrees of freedom is $n_1 - 1 + n_2 - 1$, or $20 - 1 + 20 - 1 = 38$

17. (b) $t(48) = 2.85$, $p < .01$

18. (d) No. While the new machine is significantly faster than the old machine, the effect size is too small to be meaningful.

EXERCISES

1. The eight steps to test this hypothesis would consist of the following:

    1. A statement of the null and research hypotheses:

        The null hypothesis: $H_0: \mu_1 = \mu_1$.

        The research hypothesis: $H_1: \bar{X}_1 \neq \bar{X}_2$.

2. Set the level of risk associated with the null hypothesis: .05.

3. Select the appropriate test statistic: the *t*-test for independent means.

4. Compute the test statistic value (obtained value):

$$t = \frac{\bar{X}_1 - \bar{X}_2}{\sqrt{\left[\frac{(n-1)s_1^2 + (n_2-1)s_2^2}{n_1 + n_2 - 2}\right]\left[\frac{n_1 + n_2}{n_1 n_2}\right]}} \Rightarrow$$

$$t = \frac{83.92 - 72.42}{\sqrt{\left[\frac{(12-1)10.89^2(12-1)9.49^2}{12+12-2}\right]\left[\frac{12+12}{12 \times 12}\right]}}$$

$$= \frac{11.50}{\sqrt{\left[\frac{1,304.92 + 990.92}{22}\right]\left[\frac{24}{144}\right]}}$$

$$= 2.76$$

5. Determine the value needed for rejection of the null hypothesis using the appropriate table of critical values. With degrees of freedom of 22, a two-tailed test using the .05 level of significance has a critical value of 2.074.

6. Compare the obtained value with the critical value: In this case, the obtained value is higher than the critical value.

7. and 8. Decision time: Because the obtained value is greater than the critical value, we choose to reject the null hypothesis that there is no difference between classes. In other words, there is a difference.

2. The effect size is calculated using the following equation:

$$ES = \frac{\bar{X}_1 - \bar{X}_2}{SD} \Rightarrow \frac{47.7 - 58.2}{4.3} = -2.44$$

While this effect size is a negative value, we can treat all effect sizes as positive. The effect size of 2.44 would be considered large.

3. Using this calculator, we would use 4.3 for the standard deviation of both samples. The calculated effect size (Cohen's *d*) is identical to what we had calculated by hand, −2.44.

4. First, *t* represents the test statistic that was used, 21 is the degrees of freedom, and 28.90 is the obtained value for the test statistic. Finally, $p < .05$ indicates that the probability is less than 5% that on any one test of the null hypothesis, the two groups do not differ.

SHORT-ANSWER/ESSAY QUESTIONS

1. You look for reports of effect size in each of the studies. Effect size measures the values of one group in a study relative to those of another. This indicates the amount of overlap between the values of each group. A large effect size indicates a relative lack of overlap between the two groups, which, in turn, demonstrates the practical meaningfulness of a result. Even if the authors of Article One and Article Two did not report effect sizes, in the case of *t*-tests, effect size is easy to calculate. You could do the math to see whether the studies have meaningful results, despite the large differences in sample sizes.

2. No, it is not the negative value that makes this result nonsignificant. The *t* value is small, and the *p* value is greater than .05. As with a correlation coefficient, significance is not determined by the positive or negative value of a *t* value but rather by whether the absolute value of *t* is greater than the critical value. You could also provide the mean and standard deviation of each group, which would help people to understand the data better.

3. It appears that those given red ink take longer to complete the math questions than those given black ink. The effect size is medium strength at .297, indicating that there is some meaningfulness to the difference between ink colors and the impact on math performance.

4. The requirements for a *t*-test are that you test each group only once and that there are only two groups. Since there are three groups in the anxiety and depression study, it would not be appropriate to use an independent samples *t*-test. An analysis of variance (ANOVA) would be more appropriate.

SPSS QUESTION

1. The following is the output from SPSS for this *t*-test for independent samples:

Group Statistics					
Class		N	Mean	Std. Deviation	Std. Error Mean
Exam	1.00	12	83.9167	10.89168	3.14416
	2.00	12	72.4167	9.49122	2.73988

Independent Samples Test							
	Levene's Test for Equality of Variances		*t*-test for Equality of Means				
	F	Sig.	*t*	df	Sig. (2-tailed)	Mean Difference	Std. Error Difference
Exam Equal variances assumed	.037	.850	2.757	22	.011	11.50000	4.17045
Equal variances not assumed			2.757	21.596	.012	11.50000	4.17045

These results match those that were calculated by hand. The *t* value was found to be statistically significant at the .05 level, meaning that the null hypothesis—which states that there is no difference between the two groups of respondents—should be rejected.

JUST FOR FUN/CHALLENGE YOURSELF

1. The effect size using the pooled standard deviation is calculated using the following equation:

$$ES = \frac{\overline{X_1}\overline{X_2}}{\sqrt{\frac{\sigma_1^2 + \sigma_2^2}{2}}} \Rightarrow \frac{14.2 - 17.3}{\sqrt{\frac{4.2^2 + 2.3^2}{2}}} = -.92$$

12

t(EA) for Two (Again)

Tests Between the Means of Related Groups

LEARNING OBJECTIVES

- Understand the purpose of the t-test for dependent means and when it should be used.
- Learn how to compute the observed t value for this test by hand and by using SPSS.
- Learn how to interpret the t value and understand what it means.

SUMMARY/KEY POINTS

- The t-test for dependent means is used to determine whether there is a significant difference in the scores of a group of respondents who are tested at two points in time. For example, you might want to see if neurosurgery helps to alleviate the symptoms of those prone to seizure disorders. The researcher can measure the strength of seizures prior to the treatment, perform the neurosurgery, and then measure the strength of seizures posttreatment to determine if the procedure was effective.

- The t-test for dependent means can also be used to determine whether there is a significant difference in the scores of two groups of respondents whom the researcher intentionally matched on characteristics important to the study. This study design is called matched pairs. This t-test is also known as the t-test for paired samples or the t-test for correlated samples. An industrial-organizational psychologist might be interested in looking at a new sexual harassment policy (treatment condition) versus the old policy (control condition) in a large office. Knowing that research shows that women and men may differ in their interpretation of sexual harassment, the researcher would want to make sure that men are in all conditions and women are in all conditions. Thus, he would pair up two men and send one to the experimental group and the other to the control group. He would do the same thing for the next pair of men, and the next. He would do the same for women. Thus, men and women are guaranteed to be in all study conditions. He can then determine whether the new policy is better than the old policy.

- When t-tests are used, the measure of effect size is Cohen's d:
  - A small effect size ranges from 0 to .20.
  - A medium effect size ranges from .20 to .50.
  - A large effect size is any value above .50.
  - A larger effect size represents a greater difference between the two groups.

KEY TERMS

- **t-test for dependent means:** A test of the difference between means in a group of respondents who were tested at two points in time. Also known as paired or correlated samples t-tests.

TRUE/FALSE QUESTIONS

1. The following observed value: $t_{(37)} = 5.62$ is significant at the .05 level (two-tailed test).

2. The t-test for dependent means should be used when participants are tested two or more times.

3. The effect size in a t-test for independent means is different from the effect size in a t-test for dependent means.

4. A researcher wants to measure the power of observational learning. He has some children watch an adult model aggressive behavior (hitting a Bobo doll—an inflatable doll that bounces back to a standing position when hit) versus nonaggressive behavior (playing with stickers). The researcher should use a dependent *t*-test to measure the responses of the children.

5. A researcher wants to see whether being around a gun increases levels of aggression. She angers participants in the presence of either a gun or a tennis racket. Because she thinks gun ownership may affect participants, she first asks them if they own a gun. She makes sure that some gun owners are in the gun present condition while other gun owners are in the tennis racket condition. She does the same for non-gun owners. She should use a dependent *t*-test to measure the aggressiveness of the participants.

6. A researcher measures participant attitudes about a presidential candidate before and after a debate. If he measures the same participants pre- and postdebate, he should use a dependent samples *t*-test. If he measures different participants after the debate than before the debate, he should use an independent samples *t*-test.

MULTIPLE-CHOICE QUESTIONS

1. The *t*-test for dependent means should be used in which of the following scenarios?

   a. You are comparing more than two groups that are related.

   b. You are comparing exactly two groups that are related.

   c. You are comparing exactly two groups that are unrelated.

   d. You are comparing more than two groups that are unrelated.

2. If you are running a *t*-test for dependent means on a group of 25 individuals, your degrees of freedom will be which of the following?

   a. 25

   b. 26

   c. 24

   d. 12.5

3. If you are hypothesizing that posttest scores will be lower than pretest scores, you should use a _____.

   a. one-tailed test

   b. two-tailed test

   c. *t*-test for independent means

   d. descriptive statistics

4. After conducting a *t*-test for dependent means, you arrived at the following: $t_{(22)} = 12.41$, $p < .05$. The degrees of freedom is indicated by which of the following?

   a. *t*

   b. $p < .05$

   c. 22

   d. 12.41

5. After conducting a *t*-test for dependent means, you arrived at the following: $t_{(22)} = 12.41$, $p < .05$. The test statistic is indicated by which of the following?

   a. *t*

   b. $p < .05$

   c. 22

   d. 12.41

6. After conducting a *t*-test for dependent means, you arrived at the following: $t_{(22)} = 12.41$, $p < .05$. The obtained value is indicated by which of the following?

   a. *t*

   b. $p < .05$

   c. 22

   d. 12.41

7. After conducting a *t*-test for dependent means, you arrived at the following: $t_{(22)} = 12.41$, $p < .05$. The probability is indicated by which of the following?

   a. *t*

   b. $p < .05$

   c. 22

   d. 12.41

8. An effect size of .63 would be considered _____ and _____?

   a. significant; meaningful

   b. not significant; meaningful

   c. not significant; not meaningful

   d. significant; not meaningful

9. Adam, Bob, Eric, and Jim are participating in a study. The researcher asks them for their marital status. Bob and Eric both raise their hands, indicating they are married. The researcher sends Bob to condition one and Eric to condition two. The researcher also sends Adam to condition one and Jim to condition two. What kind of statistical test should the researcher use?

   a. Independent *t*-test

   b. Dependent *t*-test for matched pairs (paired samples)

   c. Dependent *t*-test for repeated measures

   d. One-sample *z*-Test

10. You want to test children to see if their social skills improve following a film that teaches them how to make friends. You measure their social skills pre- and posttest and compare the two scores. What kind of statistical test should the researcher use?

    a. Independent *t*-test

    b. Dependent *t*-test for matched pairs (paired samples)

    c. Dependent *t*-test for repeated measures

    d. One-sample *z*-Test

11. A researcher measures the panic level of patients with arachnophobia (fear of spiders) when they see a live tarantula, which they rate at a very frightened average of 8 on a 9 point scale. He then uses exposure therapy using a virtual reality spider program to desensitize them to spiders. After the therapy, he finds that their posttreatment average level of panic when exposed to the tarantula is now a 6, $t_{(19)}$ = 2.90. Using the two-tailed test, what should the researcher conclude about the significance of the treatment?

    a. The treatment had no effect, $p > .05$.

    b. The treatment decreased level of panic, but only at $p < .05$.

    c. The treatment decreased level of panic at the $p < .01$ level.

    d. There is not enough information to make a reasonable conclusion about significance.

12. A researcher measures the panic level of patients with arachnophobia (fear of spiders) when they see a live tarantula, which they rate at a very frightened average of 8 on a 9-point scale. He then uses exposure therapy using a virtual reality spider program to desensitize them to spiders. After the therapy, he finds that their posttreatment average level of panic when exposed to the tarantula is now a 6, $t_{(19)}$ = 2.90. How many participants took part in this study?

    a. 19

    b. 20

    c. 21

    d. There is not enough information to determine how many participants there are.

13. A researcher predicts that a new antismoking campaign will reduce incidents of smoking among children. He measures smoking rates before and after the smoking campaign. What is his research hypothesis, and should he use a one-tailed or two-tailed test when comparing the obtained *t* value to the critical *t* value?

    a. The precampaign smoking rates will be higher than the postcampaign score; one-tailed test

    b. The precampaign smoking rates will be higher than the postcampaign score; two-tailed test

    c. The pre- and postcampaign smoking rates will not differ; one-tailed test

    d. The pre- and postcampaign smoking rates will not differ; two-tailed test

EXERCISE

1. Using the following data, conduct the eight steps of hypothesis testing to see whether there is a difference between these individuals' incomes before and after they return to school to get a master's degree.

Before ($1,000)	After ($1,000)
14	22
24	24
35	44
57	59
35	30
34	67
88	95
57	65

SHORT-ANSWER/ESSAY QUESTION

1. After conducting a t-test for dependent means, you found the following result: $t_{(12)} = 0.12$, $p > .05$. Interpret this result.

2. You want to look at participants' ability to complete a memory task when they are exposed to distraction (a loud noise going off in the background). First, design an independent two-group study for this experimental design. Second, design a dependent two-group study for this experimental design. Make sure to note your independent and dependent variables for each study.

3. A researcher measures intimacy ratings between unhappy married couples before giving them couple's therapy counseling (1 = not at all intimate to 10 = very intimate). After the therapy, he finds $t_{(34)} = 2.32$, $p < .05$, with the mean pretherapy ratings at 3.45 (standard deviation is 1.21) and the mean posttherapy ratings at 5.65 (standard deviation is 1.33). Interpret this result.

SPSS QUESTION

1. Input the data in the "Exercises" section into SPSS and run a t-test for dependent means. How would you interpret the results? Do your results match those calculated by hand?

TRUE/FALSE QUESTIONS

1. True.

2. False. This test should only be used when participants are tested exactly twice.

3. False. Effect size is measured in the same way for both independent and dependent means *t*-tests. The formula to determine effect size is also the same for both types of *t*-tests.

4. False. Since children watch different adult models, this study involves independent groups. The researcher should run an independent samples *t*-test.

5. True. A dependent samples *t*-test is warranted when using a matched pairs design as there is a connection between the two groups.

6. True. Measuring attitudes from the same group twice entails using a dependent *t*-test. Measuring attitudes from two different groups entails using an independent samples *t*-test.

MULTIPLE-CHOICE QUESTIONS

1. (b) You are comparing exactly two groups that are related.

2. (c) 24

3. (a) one-tailed test

4. (c) 22

5. (a) *t*

6. (d) 12.41

7. (b) $p < .05$

8. (a) significant; meaningful

9. (b) Dependent *t*-test for matched pairs (paired samples)

10. (c) Dependent *t*-test for repeated measures

11. (c) The treatment decreased level of panic at the $p < .01$ level. The critical value at the $p < .01$ level is 2.86, so the obtained value of 2.90 exceeds the critical value.

12. (b) 20, with *df* being $n - 1$, or $20 - 1 = 19$

13. (a) The precampaign smoking rates will be higher than the postcampaign score; one-tailed test

EXERCISE

1. The eight steps to test this hypothesis would consist of the following:

    1. A statement of the null and research hypotheses:
        The null hypothesis: $H_0: \mu_{pretest} = \mu_{posttest}$
        The research hypothesis: $H_1: \bar{X}_{pretest} \neq \bar{X}_{posttest}$

    2. Set the level of risk associated with the null hypothesis: .05.

3. Select the appropriate test statistic: the *t*-test for dependent means.

4. Compute the test statistic value (obtained value):

$$t = \frac{\sum D}{\sqrt{\dfrac{n\sum D^2 (\sum D)^2}{n-1}}} \Rightarrow \frac{62}{\sqrt{\dfrac{8 \times 1,376 - 3,844}{7}}} = 1.938$$

5. Determine the value needed for rejection of the null hypothesis using the appropriate table of critical values. With degrees of freedom of 7, a two-tailed test using the .05 level of significance has a critical value of 2.365.

6. Compare the obtained value with the critical value: In this case, the obtained value is lower than the critical value.

7. and 8. Decision time: Because the obtained value is lower than the critical value, we are not able to reject the null hypothesis that there is no difference between "before" and "after" values for salary. In other words, the null hypothesis is the most attractive explanation—it holds.

SHORT-ANSWER/ESSAY QUESTION

1. First, you can see that the test statistic used here was the *t*-test for dependent means. The degrees of freedom was 12, and the obtained *t* value was found to be 0.12. Finally, this result had a probability level above .05, meaning that no significant differences were found between pretest and posttest scores.

2. First, an independent groups study would assign some participants to a distraction condition and others to a no-distraction condition. The independent variable here would be type of condition, with memory performance as the dependent variable. Second, a dependent groups study would have participants first work with a distracting noise and then without a distracting noise (or vice-versa!). Using time as the independent variable (time 1 distraction session versus time 2 no-distraction session) and memory scores as the dependent variable, the study would see if memory scores changed between time 1 and time 2.

3. The couple's therapy seems to be effective, with intimacy ratings much higher posttherapy compared to pretherapy. The difference between the pretest and posttest was significant.

SPSS QUESTION

1. The following consists of the SPSS output for this test:

Paired Samples Statistics					
		Mean	N	Std. Deviation	Std. Error Mean
Pair 1	Before	43.0000	8	23.38498	8.26784
	After	50.7500	8	25.38700	8.97566

Paired Samples Correlations				
		N	Correlation	Sig.
Pair 1	Before & After	8	.896	.003

Paired Samples Test								
	Paired Differences							
		Std. Deviation	Std. Error Mean	**95% Confidence Interval of the Difference**		t	df	Sig. (2-tailed)
	Mean			Lower	Upper			
Pair 1 Before - After	−7.75000	11.31055	3.99888	217.20586	1.70586	21.938	7	.094

In this test, the two-tailed significance level is found to be .094; because this value is not below .05, we're not able to reject the null hypothesis that there is no difference between "before" and "after" scores. The calculated result for the *t* value is identical to the result calculated by hand.

13

Two Groups Too Many?

Try Analysis of Variance

LEARNING OBJECTIVES

- Understand what an analysis of variance is and when it should be used.

- Understand the difference between the *t*-test and ANOVA.

- Learn how to compute and interpret the F statistic, both by hand and by using SPSS.

- Learn how to compute η^2 (eta squared), the effect size for ANOVA.

SUMMARY/KEY POINTS

- Analysis of variance is used to test whether there is a significant difference in the mean of some dependent variable on the basis of group membership. Unlike the *t*-test, ANOVA can be used to test the difference between two groups as well as more than two groups of respondents. For example, if a researcher wants to test the facial feedback hypothesis (the idea that "wearing" an emotional expression produces an emotion congruent with that expression), they might ask some participants to put on a happy face (i.e. smile), others to put on a sad face (i.e. frown), and others to have a neutral expression. An ANOVA can determine whether happiness ratings differ among the three conditions, and it does so without the problems inherent in running a series of *t*-tests, which would increase the chances of a Type I error. That is, you would not want to run one *t*-test between sad and happy, a second *t*-test between sad and neutral, and a third *t*-test between happy and neutral, as running three different *t*-tests increases the likelihood that one of those tests would be significant only by chance. The ANOVA makes all of these comparisons within the same test.
  - The corresponding test statistic for ANOVA is the F-test.
  - The type of ANOVA covered in this chapter (the simple analysis of variance) is used for participants who are tested only once. Participants either put on a sad, happy, or neutral expression, not all three. However, a researcher *could* have a participant put on a happy face (and then measure happiness), then put on a sad face (and measure happiness a second time), and finally put on a neutral expression (and measure happiness a third time) in a repeated-measures ANOVA.

- In essence, in an ANOVA, the variance due to differences in scores is separated into variance that's due to differences between individuals within groups and variance due to differences between groups. Then, these two types of variance are compared.

- There are several types of ANOVA.
  - The simple analysis of variance, or one-way analysis of variance, includes only one factor or treatment variable in the analysis. Following are guidelines for when the one-way ANOVA is the correct statistic to use:
    - There is only one dimension or treatment. The facial expression example has one independent variable (type of facial expression).
    - There are three or more levels of the grouping factor. Our facial expression example has three levels for the single independent variable (happy, sad, and neutral), though we could easily add in additional levels (a disgust expression, a fearful expression, an angry expression, and so forth).
    - One is looking at differences across groups in average (mean) scores.
  - A factorial design includes more than one treatment factor. If you think gender might also influence ratings of happiness, you could look at males who put on either a happy, sad, or neutral face and you could have females put on a happy, sad, or neutral face. You

can then determine whether the facial expression participants put on differs between both participant gender and type of facial expressions. In this case, you would have a 3 X 2 factorial design, with three levels to the first independent variable (happy, sad, and neutral) and two levels to the second independent variable (male and female). You would need to run a factorial ANOVA for this study.

- ANOVA is an omnibus test, meaning that it tests overall differences between groups and does not tell you which groups are higher or lower than others.
  - ○ If you need to use ANOVA instead of a t-test when comparing only two groups, an F value for two groups is equal to a t value squared, or $F = t^2$. That is, you can use an ANOVA to compare the happy and sad groups only or you can use a t-test to compare these two groups. Just square the t value to find the F value, or take the square root of the F value to find the t value. Whichever you use, your obtained value must still exceed the critical tabled F value or critical tabled t value.
  - ○ Post hoc comparisons can be used to determine whether there are significant differences between specific groups. Contrast this with the comparison of means in a t-test that only looks at two conditions (e.g., happy and sad). In a t-test, you simply determine whether mean happiness ratings are higher in the sad group versus the happy group. Given a third neutral group, though, a significant ANOVA only tells you that the sad, happy, and neutral groups differ. It does not tell you if sad differs from happy, if sad differs from neutral, or if happy differs from neutral. One of the groups might differ from the other two, or all three groups might differ from each other. A post hoc test tells you which groups differ. The Bonferroni technique is an example of a frequently used post hoc test.
  - ○ For ANOVA, the measure of effect size is η^2 (eta squared). The scale of how large the effect size is as follows:
    - ○ A small effect size is about .01.
    - ○ A medium effect size is about .06.
    - ○ A large effect size is about .14.
  - ○ There are two degrees of freedom associated with the ANOVA.
    - ○ The between groups degree of freedom is the number of conditions in the independent variable minus 1, or $k - 1$.
    - ○ The within groups degree of freedom is the number of participants in the study minus the number of conditions, or $N - 1$.

KEY TERMS

- **Analysis of variance**: A test for the difference between two or more means

- **Simple analysis of variance** (aka one-way analysis of variance): A type of ANOVA in which one factor or treatment variable (such as group membership) is being explored

- **Factorial design**: A more complex type of ANOVA in which more than one treatment factor (such as group membership and gender) is being explored

- **Post hoc comparisons**: In relation to ANOVA, tests that are done in addition to ANOVA in order to look at specific group comparisons

- η^2 **(eta squared)**: The measure of effect size used for ANOVA F-tests

TRUE/FALSE QUESTIONS

1. The *F* statistic obtained from running an analysis of variance will tell you which groups have significantly higher scores as compared with every other group.

2. All *F*-tests are nondirectional.

3. A school psychologist who wants to compare reading comprehension scores between children in an advanced, regular, or remedial English class should analyze the data with an ANOVA rather than a series of *t*-tests.

4. A psychologist can use a one-way ANOVA to analyze the effectiveness of three types of therapy (psychoanalytic perspective, behavioral perspective, and cognitive perspective) and the gender of the therapist (male, female) on improving the psychological well-being of patients.

5. A researcher has participants watch one of three film commercials about the same product (a funny commercial, a sad commercial, or a drama-based commercial) and measures their willingness to buy the product. If he finds $F_{(2, 68)} = 1.23$, $p > .05$, he should run a post hoc test to see which commercial differs from the others.

MULTIPLE-CHOICE QUESTIONS

1. ANOVA is appropriate for which of the following situations?

   a. Two groups of participants are tested only once.

   b. Two groups of participants are tested twice.

   c. Three groups of participants are tested only once.

   d. Four groups of participants are tested twice.

   e. Both a and c are correct.

2. The corresponding test statistic for the ANOVA is the _____.

   a. *p* statistic

   b. *t* statistic

   c. *F* statistic

   d. *r* statistic

3. Which of the following is the type of ANOVA used when there is only one treatment factor?

   a. Simple analysis of variance

   b. Factorial design

   c. Post hoc comparisons

   d. Independent-samples *t*-test

4. Which of the following is the type of ANOVA used when there are two or more treatment factors?

   a. Simple analysis of variance

   b. Factorial design

   c. Post hoc comparisons

   d. Independent-samples *t*-test

5. Which of the following is the type of test used to look at specific group comparisons?

   a. Simple analysis of variance

   b. Factorial design

   c. Post hoc comparisons

   d. Independent-samples *t*-test

6. If you ran a factorial ANOVA using gender and social class, with the latter categorized as low, medium, or high, your factorial design would be which of the following?

   a. 2×2

   b. 3×2

   c. 1×1

   d. 1×3

7. An effect size of .12 would be considered _____.

   a. small

   b. medium

   c. large

8. An effect size of .03 would be considered _____.

   a. small

   b. medium

   c. large

9. An effect size of .16 would be considered _____.

   a. small

   b. medium

   c. large

10. If there is no difference between the distributions of scores in the groups compared within the ANOVA, your effect size will be equal to which of the following?

    a. 0

    b. 0.1

    c. 0.5

    d. 1

11. A researcher has participants who are afraid of public speaking practice a speech in one of three conditions: alone in a room with nothing on the walls; alone in a room with a mirror on the wall; or alone in a room with a video camera that appears to be taping them. He later has them give their speech to a live audience, and he measures their level of nervousness. He finds $F_{(2, 100)} = 4.65$, $p < .05$. How many participants are in the study?

    a. 2

    b. 100

    c. 102

    d. 103

12. A researcher studying decision-making assigns some participants to work in a group with a cooperative agenda, some to work in a group with a competitive agenda, and some to work alone. He then measures their satisfaction with their work and finds $F_{(2, 65)} = 5.95$. What should the researcher conclude?

    a. There is a significant difference in satisfaction among the three groups at the $p < .05$ level but not the $p < .01$ level.

    b. There is a significant difference in satisfaction among the three groups at both the $p < .05$ level and the $p < .01$ level.

    c. There is no significant difference in satisfaction among the three groups at either the $p < .05$ level or the $p < .01$ level.

    d. The researcher cannot draw a conclusion at all in this study.

13. A researcher asks religious students at a seminary school to give a lecture on the Good Samaritan Bible parable, a story in which a good Samaritan provides aid to a man who was robbed and beaten. Noting that the lecture hall is on the other side of campus, the researcher tells the students one of three things: "You are late and need to hurry across campus to give the lecture"; "You will be right on time if you leave now"; or "You have a lot of time to get to the other side of campus." Along the route to the lecture hall, the students pass a man who is clearly in need of help. The researcher predicts that those in no rush will help the man while those right on time or rushed will pass by without providing help. What is the best statistical test to use when analyzing this study?

    a. A t-test for independent samples

    b. A simple (one way) analysis of variance

    c. A repeated measures analysis of variance

    d. None of the above

14. Reconsider the Good Samaritan question above. If the researcher measures how quickly the participants provide help, which of the following should he use to analyze the time to help dependent variable?

    a. A t-test for independent samples

    b. A simple (one way) analysis of variance

    c. A repeated measures analysis of variance

    d. None of the above

15. A researcher finds significance in a study looking at the effect of text color on reading ability, $F_{(4, 78)} = 3.10, p > .05$. How many groups are in this study and how many participants are there?

    a. 4 color groups; 78 participants

    b. 4 color groups; 83 participants

    c. 5 color groups; 78 participants

    d. 5 color groups; 83 participants

    e. None of the above

EXERCISES

1. If you have MS between groups of 2.4 and MS within groups of 0.3, what would your calculated F statistic be?

2. You are interested in seeing how self-conscious Caucasian participants feel when serving as jurors in a criminal trial in which the defendant is African American. You assign Caucasian participants to juries in which all other jurors are Caucasian, all other jurors are African American, or there is a mix of Caucasian and African American jurors. The self-consciousness score is calculated on a scale of zero to 100, with 100 indicating the highest level of self-consciousness. Using the following data, conduct the eight steps of hypothesis testing.

All Caucasian Jury	Mixed Jury	All African American Jury
23	47	88
43	77	98
56	84	78
89	55	76
45	67	82
55	76	95
23	45	79
33	67	85
27	87	94
26	66	87

3. Using the results from the previous question, construct an F table (an example is presented on page 252 of the text).

SHORT-ANSWER/ESSAY QUESTIONS

1. What is the critical F statistic value if your total sample size is 50 and you are comparing three groups of respondents (at the .05 level of significance)?

2. What is the critical F statistic value at the .05 level of significance for the following: $F_{(4, 70)}$?

3. How would you interpret $F_{(2, 30)} = 32.60$, p < .05?

4. A researcher finds that participants exposed to high, medium, or low levels of stress (as manipulated by the amount of time they have to complete a task) differ in their ratings of personal anxiety. Why would you not want to use t-tests to compare the high and medium conditions, the high and low conditions, and the medium and low conditions? What would you use instead?

5. Imagine you are conducting a study looking at gender role stereotypes and hiring decisions. Your independent variable is the job résumé of four individuals: a male, a female, a transgender male, and a transgender female. Your dependent variable is likelihood of hiring the individual. Using these independent and dependent variables, first design a study in which your analysis would involve a simple ANOVA with independent means. Second, design a study in which your analysis would involve a repeated-measures ANOVA (dependent means).

6. In a study looking at participants' likelihood of conforming to the majority opinion in a decision-making task (as manipulated by requiring group unanimity before the experiment can end, requiring a simple majority consensus before the experiment can end, or not requiring consensus before the experiment can end), the researcher runs a simple ANOVA and finds $F_{(2, 40)} = 5.00$. First, how many participants are in this study? Second, what should the researcher conclude?

SPSS QUESTION

1. Run an analysis of variance using the data in the "Exercises" section, question 2 (additionally, run the Bonferroni post hoc comparison). Do your results match those calculated by hand? How would you interpret these results?

JUST FOR FUN/CHALLENGE YOURSELF

1. If you are only comparing two groups of respondents, and the t value is found to be 2.30, what will the F statistic be?

2. If someone performs multiple t-tests, and the initial Type I error rate is .05 and 8 comparisons are made, what will the actual Type I error rate be?

TRUE/FALSE QUESTIONS

1. False. Post hoc comparisons are necessary for this—the F statistic is nondirectional.

2. True.

3. True. Running multiple t-tests for all of the potential comparisons (remedial versus advanced, remedial versus regular, and advanced versus regular) would increase the chances of making a Type I error.

4. False. A one-way ANOVA focuses on one independent variable only. There are two independent variables in this design (type of therapy and therapist gender).

5. False. Post hoc tests are only necessary when the F-test is significant.

MULTIPLE-CHOICE QUESTIONS

1. (c) Three groups of participants are tested only once.

2. (c) F statistic

3. (a) Simple analysis of variance

4. (b) Factorial design

5. (c) Post hoc comparisons

6. (b) 3×2

7. (b) medium

8. (a) small

9. (c) large

10. (a) 0

11. (d) 103. To determine the number of participants, remember that the df is $N - k$, or the number of participants (N) minus the number of groups (k). Since df is 100, this means N must be 103, or $103 - 3 = 100$.

12. (b) There is a significant difference in satisfaction among the three groups at both the $p < .05$ level and the $p < .01$ level.

13. (d) None of the above. Since the dependent variable is based on a yes or no response, you cannot use the mean. Inferential tests like the t-test and ANOVA require that the dependent variable is interval or ratio (a scaled variable) rather than a nominal or categorical response like yes or no, and thus neither the t-test nor ANOVA is appropriate here.

14. (b) A simple (one way) analysis of variance since there are three conditions (late, on time, lots of time) and a scaled dependent variable

15. (d) 5 color groups; 83 participants. Consider $F_{(4, 78)}$. The between groups degrees of freedom is the number of groups minus 1, or $k - 1$. Here, k is 5, thus $5 - 1 = 4$. The within groups degrees of freedom is the number of participants minus the number of groups, or $N - k$. Here, N is 83 and k is 5, thus $83 - 5 = 78$. Consequently, $F_{(4, 78)}$ has 5 color groups and 83 participants.

1. This would be calculated in the following way:

$$F = \frac{MS_{between}}{MS_{within}} = \frac{2.4}{0.3} = 8.0$$

2. The eight steps of hypothesis testing:

   1. State the null and research hypotheses:

      $H_0: \mu_1 = \mu_2 = \mu_3$

      $H_1: \bar{X}_1 \neq \bar{X}_2 \neq \bar{X}_3$

   2. Set the level of significance: 0.05.

   3. Select the appropriate test statistic: a simple ANOVA.

   4. Compute the test statistic value:

	Office Workers	**Students**	**Musicians**
n	10	10	10
ΣX	420	671	862
$\Sigma(X^2)$	21,508	46,923	74,828
$(\Sigma X)^2/n$	17,640.0	45,024.1	74,304.4

$\Sigma\Sigma X = 1,953.$

$(\Sigma\Sigma X)^2/N = 127,140.3.$

$\Sigma\Sigma(X^2) = 143,259.$

$\Sigma(\Sigma X)^2/n = 139,968.5.$

$SS_{Between} = \Sigma(\Sigma X)^2/n - (\Sigma\Sigma X)^2/N = 139,968.5 - 127,140.3 = 9.828.2.$

$SS_{Within} = \Sigma\Sigma(X^2) - \Sigma(\Sigma X)^2/n = 143,259 - 139,968.5 = 6,290.5.$

$MS_{Between} = SS_{Between}/(k-1) = 9.828.2/2 = 4914.1.$

$MS_{Within} = SS_{Within}/(N-k) = 6,290.5/(30 - 3) = 232.98.$

$F = MS_{Between}/MS_{Within} = 4914.1/232.98 = 21.09223.$

   5. Determine the value needed to reject the null hypothesis: Critical $F_{(2, 27)} = 3.36$.

   6. Compare the obtained value with the critical value: The obtained value, 21.09, is larger than the critical value of 3.36.

   7. and 8. Decision time: Because the obtained value is greater than the critical value, we would reject the null hypothesis that states there is no difference between these groups.

3. The F table:

Source	**Sum of Squares**	**df**	**Mean Sum of Squares**	**F**
Between groups	9,828.2	2	4,914.1	21.09
Within groups	6,290.5	27	232.98	
Total	16,118.7	29		

SHORT-ANSWER/ESSAY QUESTIONS

1. The critical F statistic in this case would be 3.21. This result is obtained by being more conservative and looking at the value that corresponds to a degrees of freedom of 45.

2. The critical F statistic in this case is 2.51.

3. First, the F represents the test statistic that was used. Then, 2 and 30 represent the degrees of freedom for the between-group and within-group estimates, respectively. The value of 32.60 represents the obtained value, which was arrived at by using the formula for the F statistic. Finally, $p < .05$ indicates that the probability is less than 5% that the average scores between-groups differ due to chance as opposed to the effect of the treatment or group membership. This also indicates that the research hypothesis should be preferred over the null hypothesis, as there is a significant difference.

4. Running multiple t-tests (like high versus medium, low versus medium, and low versus high) increases the Type I error, or finding a significant difference when there really is no difference. The better method is to run a simple ANOVA. If significant, post hoc tests can determine which of the three conditions differ.

5. In a simple ANOVA for independent means, the researcher would randomly assign participants to view one of the four résumés (male, female, transgender male, and transgender female) and rate their likelihood of hiring that one individual. In the repeated measures ANOVA for dependent means, the participant would look at all four résumés and provide their likelihood of hiring ratings for each.

6. First, there are 43 participants in this study. The within degrees of freedom relies on the number of participants minus the number of groups, or $N - k$. Here, 43 participants minus 3 groups equals a df of 40. Second, the researcher should conclude that the three conditions do not differ. The obtained F value of 5.00 does not exceed the critical value of 5.18 required for df 2, 40.

SPSS QUESTION

1. The SPSS output is shown here:

One-Way

ANOVA

Source	Sum of Squares	df	Mean Squares	F	Sig.
Between groups	9,828.200	2	4,914.100	21.092	.000
Within groups	6,290.500	27	232.981		
Total	16,118.700	29			

Post Hoc Tests

MULTIPLE COMPARISONS

Score

Bonferroni

(I) Group	(J) Group		Mean Difference (I – J)	Std. Error	Sig.	95% Confidence Interval	
						Lower Bound	Upper Bound
dimension2	1 dimension3	2	−25.100*	6.826	.003	−42.52	−7.68
		3	−44.200*	6.826	.000	−61.62	−26.78
	2 dimension3	1	25.100*	6.826	.003	7.68	42.52
		3	−19.100*	6.826	.028	−36.52	−1.68
	3 dimension3	1	44.200*	6.826	.000	26.78	61.62
		2	19.100*	6.826	.028	1.68	36.52

* The mean difference is significant at the 0.05 level.

These results, in regard to the one-way ANOVA, do match those calculated by hand. First, the F-test was found to be statistically significant, indicating that the null hypothesis should be rejected in favor of the research hypothesis. In other words, there is a significant difference between groups. Additionally, the Bonferroni post hoc comparison finds a significant difference among all three groups of respondents. Focusing on the final row, which compares group 3 (musicians) against groups 1 (office workers) and 2 (students), and looking at the mean difference and significance columns, we can see that musicians are significantly happier than both students and office workers (no surprise). Additionally, the only other comparison, which was between groups 1 and 2, is shown in the first row of results. Here, we can see that office workers are significantly less happy than are students (again, no surprise here).

JUST FOR FUN/CHALLENGE YOURSELF

1. To obtain the F statistic in this case, we simply need to square the t value: This gives us a calculated F statistic of 5.29.

2. If someone performs multiple t-tests, and the initial Type I error rate is .05 and 8 comparisons are made, what will the actual Type I error rate be? This can be calculated by using the following equation:

$$\text{True Type I error} = 1 - (1 - \alpha)^k = 1 - (1 - .05)^8 = 0.34$$

14

Two Too Many Factors

Factorial Analysis of Variance: A Brief Introduction

LEARNING OBJECTIVES

- Learn when it is appropriate to use the factorial analysis of variance.

- Understand the distinction between main effects and interaction effects and what they indicate.

- Learn how to use SPSS to conduct a factorial analysis of variance.

- Computing the effect size for factorial analysis of variance.

SUMMARY/KEY POINTS

- The factorial analysis of variance, or two-way analysis of variance, is used when you have more than one factor. Factors are also referred to as independent or treatment variables. For example, imagine that you want to test the theory of learned helplessness in which repeated failure leads people to stop trying. Let's say you give them the word CINERAMA and ask them to turn it into a new word using the same letters, and you time how quickly they come up with the word AMERICAN. However, before giving them the word CINERAMA, you give all participants practice words to rearrange, with some participants getting words that are easy to unscramble (words like TAB and STEW, which can be rearranged to spell the BAT and WEST, respectively), while others get practice words that are impossible to rearrange (words ORANGE and WHIRL, which cannot be altered). This two-condition study might show you that those given the impossible-to-rearrange practice words take significantly longer to unscramble the word CINERAMA than those given the easy-to-unscramble practice words, as the former group's repeated failure unscrambling the practice words leads them to stop trying to solve a word like CINERAMA, which is actually solvable. Yet you might wonder whether participant gender makes a difference. That is, will men or women take longer to unscramble the word CINERAMA, and will they do so differently in the easy-to-solve versus impossible-to-solve practice word conditions? A factorial ANOVA can look at the differences between these two different independent variables (practice word difficulty and participant gender).
  - This type of ANOVA can test the significance of the main effects of each independent variable, as well as the significance of the interaction effect between independent variables. That is, you can see if those in the easy practice word condition unscramble the word CINERAMA quicker than those in the impossible practice word condition for one main effect. For a second main effect, you can see if male participants unscramble the word CINERAMA faster than female participants. Finally, you can look at the interaction of practice word condition and gender to see who unscrambled the word CINERAMA faster: males in the easy condition, males in the impossible condition, females in the easy condition, or females in the impossible condition.
  - This type of ANOVA is used when participants are tested only once. In the learned helplessness study, we have four total conditions, with participants assigned to only one of them (male easy, male impossible, female easy, and female impossible).
  - The test statistic used is the factorial analysis of variance.
  - There will be at least three hypotheses: a main effect for the first factor, a main effect for the second factor, and an interaction. There will similarly be three F-tests for each component: an F-test for the first main effect; an F-test for the second main effect; and an F-test for the interaction.
  - Factorial ANOVA allows for more complex research questions to be asked. Researchers often want to know if responses differ with a combination of two independent variables. As the real word is itself very complex, more complex research designs better

mimic the real world, increasing the applicability of the research study beyond the laboratory setting. Of course, research designs can add more than two levels to each independent variable, such as a third level of practice (a moderate condition might include both easy practice words like TAB as well as impossible practice words like WHIRL) and the gender independent variable, creating a 3 × 2 design. Perhaps the researcher also wants to know if the participant suffers from clinical depression (yes or no) to see how this relates to their ability to unscramble words, creating a 3 × 2 × 2 design.

○ For factorial ANOVA, effect size is measured by ω^2 (omega squared), and the point is still to make a judgment about the magnitude of an observed difference.

KEY TERMS

- **Factorial analysis of variance** (aka two-way analysis of variance): The type of ANOVA that is used when there is more than one independent variable or factor

- **Main effect**: In analysis of variance, a significant effect of a factor, or independent variable, on the outcome variable

- **Source table**: A listing of sources of variance in an analysis of variance summary table. This will be produced in SPSS output and will show the obtained F value and level of significance for each factor and the interaction, as well as other information.

- **Interaction effect**: The varying effect of one independent variable on the dependent variable depending on the level of a second independent variable. For example, a significant interaction between exercise treatment (low- or high-impact exercise) and gender indicates that females lose more weight than males in the high-impact treatment condition and males lose more weight than females under the low-impact condition.

- ω^2 (**omega squared**): The name of the effect size used for factorial ANOVA

TRUE/FALSE QUESTIONS

1. A factorial analysis of variance can be used only in the case where you have two independent, or treatment, variables.

2. Interaction effects are always significant when all main effects are significant.

3. If the authors of an article you read do not provide the value of ω^2 but do provide a full source table/F table, you can calculate the effect size yourself.

4. In order to test the idea of contact theory in reducing whether people stereotype minority members, a researcher has some participants work with minority members who behave in a stereotype-confirming manner while other participants work with minority members who behave in a stereotype-disconfirming manner. The researcher should use a factorial ANOVA to analyze the extent to which participants maintain their stereotypes after their interactions.

5. A factorial design can look at more than two levels for each independent level.

6. A researcher is interested in looking at frustration. He assigns participants to unscramble easy words versus hard words, and he measures their level of depression (high versus low). He finds significant differences for those solving hard words versus easy words. He does not find differences between those who are depressed and those who are not depressed. However, he does find that depressed participants who solved hard words

rated their level of depression higher than all other conditions, though the other conditions did not differ. In this example, there are two significant main effects and a significant interaction.

MULTIPLE-CHOICE QUESTIONS

1. A factorial analysis of variance should be used in which of the following situations?

    a. You have two independent variables, and participants are tested more than once.

    b. You have one independent variable, and participants are tested more than once.

    c. You have two independent variables, and participants are tested only once.

    d. You have three independent variables, and participants are tested more than once.

2. Which of the following statements is true about a factorial analysis of variance?

    a. The number of main effects will be equal to the number of independent variables.

    b. The number of main effects will be equal to the number of independent variables plus the number of interaction effects.

    c. The number of main effects will be equal to the number of interaction effects.

    d. The number of main effects will be equal to the number of independent variables minus the number of interaction effects.

3. What does the following constitute: "The effect of being upper-class, middle-class, or lower-class is different for males and females"?

    a. An interaction effect

    b. A main effect

    c. Either an interaction effect or a main effect

    d. None of the above

4. When plotted on a graph, a significant interaction effect is indicated by _____.

    a. parallel lines

    b. lines that cross/are not parallel

    c. either a or b

    d. none of the above

5. A researcher has student participants read and rate the persuasiveness of a campus newspaper article about raising tuition at a local university to cover costs of upgrading the library. The article is either for or against the tuition increase, and the article is attributed to either a student, a professor, or a dean at the university. What kind of factorial design is this?

    a   A 1 × 2 factorial design

    b. A 2 × 2 factorial design

    c. A 2 × 3 factorial design

    d. It is not a factorial design

6. Stanley Schracter designed a study where he gave participants either a placebo or a drug that increased their heart rate. He told some participants that the drug had no effect while others were told the drug increased heart rate. Finally, he placed some in a room with a confederate who acted either very angry or very happy. How many independent variables are there in this study, and how many levels are there for each independent variable?

   a. Two independent variables with two levels each

   b. Two independent variables with three levels each

   c. Three independent variables with two levels each

   d. Three independent variables with three levels each

7. A researcher wants to looks at the attractiveness level of a potential new employee (attractive versus unattractive) and experience level of the potential new employee (five years versus ten years) to determine which characteristic is most important in ratings of whether she should be hired. The analysis might result in all of the following EXCEPT:

   a. No significant main effects and no significant interaction

   b. One significant main effect, one nonsignificant main effect, and a significant interaction

   c. No significant main effects and a significant interaction

   d. Two significant main effects and a significant interaction

   e. All of the above are possible

8. A researcher wants to test the effectiveness of hypnosis versus meditation on calming those who rate themselves as either highly anxious or moderately anxious. Which of the following represents the null hypothesis for the interaction?

   a. H_0: $\mu_{hypnosis} = \mu_{meditation}$

   b. H_0: $\mu_{high} = \mu_{moderate}$

   c. H_0: $\mu_{hypnosis-high} = \mu_{meditation-moderate} = \mu_{meditation-high} = \mu_{meditation-moderate}$

   d. H_0: $\mu_{hypnosis-high} \neq \mu_{meditation-moderate} \neq \mu_{meditation-high} \neq \mu_{meditation-moderate}$

EXERCISE

1. Draw an example of a graph that illustrates a strong interaction effect. Now, draw a graph that illustrates no interaction effect at all.

SHORT-ANSWER/ESSAY QUESTION

1. In what situations would a factorial ANOVA be preferred over the one-way ANOVA and why?

2. Inappropriate student/teacher conduct has been in the news a lot lately. Imagine a researcher wants to determine whether observers find such conduct more or less inappropriate depending on whether the teacher is male or female (teacher gender) and whether the student is male or female (student gender). Identify which conditions are present in each main effect and which conditions are involved in the interaction.

3. Imagine you get the following ANOVA source table for question number two above. Interpret the table and write up the main effects and interaction as you would see it in a journal article:

Tests of Between-Subjects Effects Dependent Variable: Impropriety Score

Source	Type III Sum of Squares	df	Mean Square	F	Stg.	Partial Eta Squared
Corrected Model	126.700[a]	3	42.233	19.393	.000	.618
Intercept	2160.900	1	2160.900	992.250	.000	.965
Teacher Gender	96.100	1	96.100	44.128	.000	.551
Student Gender	8.100	1	8.100	3.719	.062	.094
Teacher Gender* Student Gender	22.500	1	22.500	10.332	.003	.223
Error	78.400	36	2.178			
Total	2366.00	40				
Corrected Total	205.100	39				

[a] R Squared = .618 (Adjusted R Squared = .586)

SPSS QUESTION

1. Using the following data, use the eight steps of hypothesis testing to determine the main effects of group membership and the interaction between independent variables. Additionally, write up the three results as they would be printed in a journal article or report (see page 272 of the text for an example). The dependent variable is health scores, while the independent variables are gender and social class. With regard to health scores, higher values indicate better overall physical health.

Note: Include both independent variables as fixed factors.

Health Score	Gender	Social Class
98	Female	Upper
88	Male	Upper
87	Female	Upper
85	Female	Middle
75	Female	Upper
74	Male	Upper
72	Male	Middle
71	Female	Middle
65	Male	Middle
55	Male	Upper
47	Male	Middle
33	Male	Lower
22	Female	Lower
10	Male	Upper
5	Male	Lower

JUST FOR FUN/CHALLENGE YOURSELF

1. Using the following data, conduct a multivariate analysis of variance in SPSS and interpret the results. The two dependent variables are health scores and lifestyle attitudes. Higher values on the health score variable indicate better general physical health, while higher scores on the lifestyle attitudes variable indicate more positive and more healthy attitudes.

Lifestyle Attitudes	Health Score	Gender	Social Class
88	98	Female	Upper
76	88	Male	Upper
98	87	Female	Upper
76	85	Female	Middle
67	75	Female	Upper
86	74	Male	Upper
65	72	Male	Middle
67	71	Female	Middle
65	65	Male	Middle
43	55	Male	Upper
54	47	Male	Middle
22	33	Male	Lower
32	22	Female	Lower
11	10	Male	Upper
2	5	Male	Lower

ANSWER KEY

TRUE/FALSE QUESTIONS

1. False. The factorial analysis of variance can be used when you have two or more than two independent/treatment variables.

2. False. Whether the interaction effects are significant will not depend on the significance of the main effects.

3. True. To calculate ω^2, you need $SS_{between}$, $df_{between}$, MS_{within}, and SS_{total}, all of which can be found in the source table of a factorial ANOVA F-test.

4. False. This describes a two group design (stereotype-confirming versus stereotype-disconfirming), and thus a t-test is more appropriate.

5. True. Factorial designs can have an unlimited number of independent variables as well as an unlimited number of levels for those independent variables, though there is a practical limit to designs. Going beyond four or five independent variables gets very complicated, time consuming, and requires a lot of participants.

6. False. There is a main effect for the difficulty of the word (hard differs from easy), but no main effect for depression (depressed versus not depressed). There is also a significant interaction of word difficulty and depression.

MULTIPLE-CHOICE QUESTIONS

1. (c) You have two independent variables, and participants are tested only once.

2. (a) The number of main effects will be equal to the number of independent variables.

3. (a) An interaction effect

4. (b) lines that cross/are not parallel

5. (c) A 2 × 3 factorial design (2 levels of argument: for or against the tuition raise; 3 levels of author: student, professor, or dean)

6. (c) Three independent variables with two levels each

7. (e) All of the above are possible

8. (c) H_0: $\mu_{hypnosis-high} = \mu_{meditation-moderate} = \mu_{meditation-high} = \mu_{meditation-moderate}$

EXERCISE

1. This first graph presents an example of a strong interaction effect; the lines cross and have very different slopes.

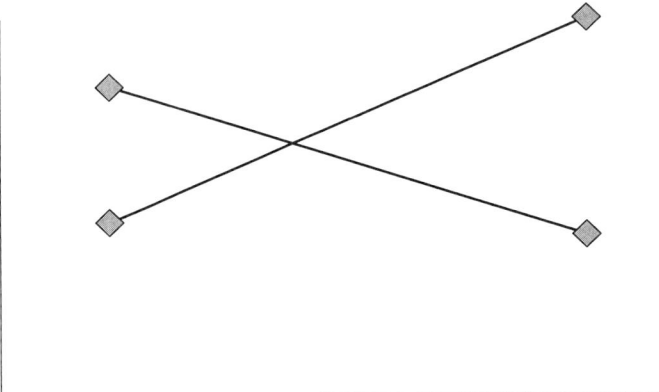

This next graph illustrates no interaction effect, as the lines are parallel to each other.

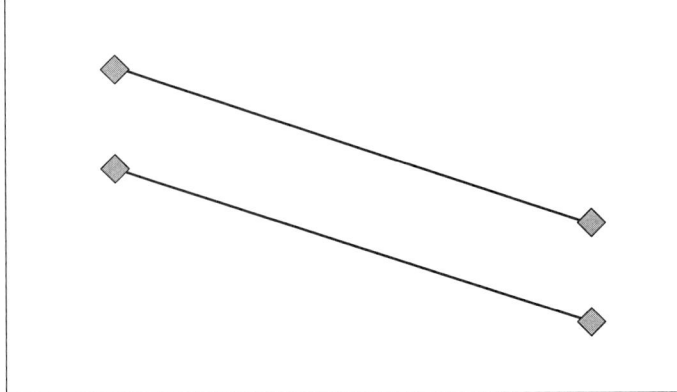

SHORT-ANSWER/ESSAY QUESTION

1. A factorial ANOVA is preferred over a one-way ANOVA whenever there is more than one independent, or treatment, variable. The factorial ANOVA is preferred because a one-way ANOVA can only incorporate a single independent variable into the analysis.

2. The main effect for teacher gender involves male versus female teachers. The main effect for student gender involves male versus female students. The interaction involves male teachers and male students, male teachers and female students, female teachers and male students, and female teachers and female students.

3. There is a main effect for teacher gender, $F_{(1, 36)} = 44.13$, $p < .001$. There is no main effect for student gender, $F_{(1, 36)} = 3.72$, $p > .05$. There is an interaction for teacher gender by student gender, $F_{(1, 36)} = 10.33$, $p < .01$

SPSS QUESTIONS

1. The eight steps of hypothesis testing:

    1. State the null and research hypotheses:

        The null hypotheses:

        For gender: H_0: $\mu_{male} = \mu_{female}$

        For social class: H_0: $\mu_{upper} = \mu_{middle} - \mu_{lower}$

        For the interaction effect:

        H_0: $\mu_{upper \bullet male} = \mu_{upper \bullet female} = \mu_{middle \bullet male} = \mu_{middle \bullet female} = \mu_{lower \bullet male} = \mu_{lower \bullet female}$

        The research hypotheses:

        For gender: H_1: $\bar{X}_{male} \neq \bar{X}_{female}$

        For social class: H_1: $\bar{X}_{upper} \neq \bar{X}_{middle} \neq \bar{X}_{lower}$

        For the interaction effect:

        $\bar{X}_{upper \bullet male} \neq \bar{X}_{upper \bullet female} \neq \bar{X}_{middle \bullet male} \neq \bar{X}_{middle \bullet female} \neq \bar{X}_{lower \bullet male} \neq \bar{X}_{lower \bullet female}$

    2. Set the level of significance: 0.05.

    3. Select the appropriate test statistic: a factorial ANOVA.

    4. Compute the test statistic value. The SPSS output relating to this test is presented in the following tables:

Between-Subjects Factors

		N
Gender	Female	6
	Male	9
Social_Class	Lower	3
	Middle	5
	Upper	7

Tests of Between-Subjects Effects Dependent Variable: Health_Score

Source	Type III Sum of Squares	df	Mean Square	F	Sig.
Corrected Model	7,623.650[a]	5	1,524.730	3.016	.072
Intercept	35,936.250	1	35,936.250	71.081	.000
Gender	842.917	1	842.917	1.667	.229
Social_Class	5,517.382	2	2,758.691	5.457	.028
Gender * Social_Class	373.808	2	186.904	.370	.701
Error	4,550.083	9	505.565		
Total	64,625.000	15			
Corrected Total	12,173.733	14			

[a] R squared = .626 (adjusted R squared = .419).

5. Determine the value needed for rejection of the null hypothesis: SPSS automatically performs this step, so it does not need to be done manually.

6. Compare the obtained value and the critical value: This step can be completed by looking at the significance values for the main and interaction effects presented in the previous table. As we can see, the main effect of gender is not significant, while the main effect of social class is. Additionally, the interaction between gender and social class is not significant.

7. and 8. Decision time: The null hypothesis that there is no difference in health scores on the basis of gender cannot be rejected, because the main effect is not significant. Additionally, the null hypothesis that there is no interaction effect between gender and social class cannot be rejected, because this interaction effect is also not significant. However, the null hypothesis suggesting no difference in health scores on the basis of social class can be rejected, because this main effect was found to be significant. This suggests that while there is no difference in health scores on the basis of gender and there is no interaction between gender and social class, there is a significant difference in health scores on the basis of social class.

These three results could be written up in the following way, as might be seen in a journal article:

Gender: $F_{(1,9)} = 1.67$, $p = .229$.

Social class: $F_{(2,9)} = 5.46$, $p < .05$.

Interaction between gender and social class: $F_{(2,9)} = 0.37$, $p = .701$.

JUST FOR FUN/CHALLENGE YOURSELF

1. The results are presented in the following tables:

Between-Subjects Factors

		N
Gender	Female	6
	Male	9
Social_Class	Lower	3
	Middle	5
	Upper	7

Multivariate Tests

Effect		Value	F	Hypothesis df	Error df	Sig.
Intercept	Pillai's trace	.892	33.155[a]	2.000	8.000	.000
	Wilks's lambda	.108	33.155[a]	2.000	8.000	.000
	Hotelling's trace	8.289	33.155[a]	2.000	8.000	.000
	Roy's largest root	8.289	33.155[a]	2.000	8.000	.000
Gender	Pillai's trace	.238	1.251[a]	2.000	8.000	.337
	Wilks's lambda	.762	1.251[a]	2.000	8.000	.337
	Hotelling's trace	.313	1.251[a]	2.000	8.000	.337
	Roy's largest root	.313	1.251[a]	2.000	8.000	.337
Social_Class	Pillai's trace	.552	1.717	4.000	18.000	.190
	Wilks's lambda	.449	1.969[a]	4.000	16.000	.148
	Hotelling's trace	1.223	2.140	4.000	14.000	.130
	Roy's largest root	1.220	5.490[b]	2.000	9.000	.028
Gender * Social_Class	Pillai's trace	.286	.751	4.000	18.000	.570
	Wilks's lambda	.727	.690[a]	4.000	16.000	.610
	Hotelling's trace	.356	.623	4.000	14.000	.654
	Roy's largest root	.293	1.319[b]	2.000	9.000	.314

[a] Exact statistic
[b] The statistic is an upper bound on F that yields a lower bound on the significance level.
[c] Design: Intercept + Gender + Social_Class + Gender * Social_Class

Tests of Between-Subjects Effects

Source	Dependent Variable	Type III Sum of Squares	df	Mean Square	F	Sig.
Corrected Model	Lifestyle_Attitudes	7428.567[a]	5	1485.713	3.110	.066
	Health_Score	7623.650[b]	5	1524.730	3.016	.072
Intercept	Lifestyle_Attitudes	34056.010	1	34056.010	71.283	.000
	Health_Score	35936.250	1	35936.250	71.081	.000
Gender	Lifestyle_Attitudes	1254.943	1	1254.943	2.627	.140
	Health_Score	842.917	1	842.917	1.667	.229
Social_Class	Lifestyle_Attitudes	4620.377	2	2310.188	4.835	.037
	Health_Score	5517.382	2	2758.691	5.457	.028
Gender* Social_Class	Lifestyle_Attitudes	289.313	2	144.656	.303	.746
	Health_Score	373.808	2	186.904	.370	.701
Error	Lifestyle_Attitudes	4299.833	9	477.759		
	Health_Score	4550.083	9	505.565		
Total	Lifestyle_Attitudes	60122.000	15			
	Health_Score	64625.000	15			
Corrected Total	Lifestyle_Attitudes	11728.400	14			
	Health_Score	12173.733	14			

[a] R Squared = .633 (Adjusted R Squared = .430)
[b] R Squared = .626 (Adjusted R Squared = .419)

Based on the final table, we see that these results suggest that social class is a significant predictor of both health scores and lifestyle attitudes, while neither gender nor the interaction between gender and social class significantly predicts either of these two dependent variables.

15 Cousins or Just Good Friends?

Testing Relationships Using the Correlation Coefficient

LEARNING OBJECTIVES

- Learn how to test for the significance of a correlation coefficient and how to interpret the results.

- Review the difference between significance and causality in relation to correlation coefficients.

- Review the difference between significance and meaningfulness.

- Learn how to use SPSS to calculate the significance of a correlation coefficient.

SUMMARY/KEY POINTS

- This chapter covers correlation coefficients, which were discussed earlier in the text, but this chapter also covers the use of statistical significance in relation to correlation coefficients.
  - Correlation coefficients examine the relationship between variables, not the difference between groups, and can be either direct (positive) or indirect (negative). For example, a researcher might find that the more tolerant a workplace is of sexual harassment, the more sexual harassment there will be at the workplace. That is, he finds a direct correlation in which increased tolerance correlates with increased incidents of sexual harassment (or decreased tolerance correlates with decreased incidents of sexual harassment). An indirect correlation might involve more sexual harassment training seminars correlating with fewer incidents of sexual harassment at work.
  - A correlation coefficient can only test two variables at a time.
  - The appropriate test statistic to use is the *t*-test for the correlation coefficient.
  - Tests can be either directional or nondirectional. Use a directional (or one-tailed) test when you hypothesize that the relationship will be either positive or negative. If you think increased tolerance correlates with increased incidents of sexual harassment, use a one-tailed test. If you think decreased tolerance correlates with decreased incidents of sexual harassment, use … a one-tailed test! Use a one-tailed test even if you think the variables will go in opposite directions (increasing training decreases incidents). You will only use a two-tailed test if you do not predict a direction for the correlation (that is, if increasing tolerance will correlate with sexual harassment incidents in some unspecified direction).

- A significant correlation does not indicate causality. Just because a workplace is tolerant of sexual harassment doesn't mean the workplace *causes* harassment, just like the direct correlation between consuming ice cream does not *cause* higher levels of crime.

- A significant correlation does not necessarily indicate a meaningful relationship. A more important determinant of meaningfulness is how large the coefficient of determination is and how small the coefficient of alienation is. The coefficient of determination indicates how much of the variance in one variable is accounted for by the variance in the other variable. For example, tolerance of sexual harassment might be a big factor in the number of incidents of sexual harassment in the workplace, but there are other factors that might account for such incidents, including the number of men and women in the office; whether the work is in a traditionally male setting (like construction) or traditionally female setting (like nursing); whether the head of the company is male or female, etc. If tolerance is the primary factor related to incidents of sexual harassment, tolerance will have a large coefficient of determination. If tolerance is one of many factors, the coefficient of alienation will be high.

TRUE/FALSE QUESTIONS

1. The correlation coefficient can only be used for two-tailed tests.

2. A significant correlation between two variables does not imply that one variable causes the other.

3. A significant correlation indicates a meaningful relationship exists between the two variables in the analysis.

4. A researcher looking at the correlation between participant age and authoritarianism finds $r_{(25)} = .67$, $p < .05$, indicating that being older causes people to be more authoritarian.

5. A large coefficient of alienation between increased aggression and increased temperature indicates a large overlap between these two variables.

6. It is possible for a researcher to find a correlation of $r_{(34)} = 2.34$, $p < .05$ between the amount of time students take to graduate from college and their salary during their first year postgraduation.

MULTIPLE-CHOICE QUESTIONS

1. The correlation coefficient examines _____.

   a. differences between two groups

   b. differences between two or more groups

   c. the relationship between two variables

   d. the relationship between two or more variables

2. In the case of the correlation coefficient, the appropriate test statistic to use is the _____.

   a. F-test for the correlation coefficient

   b. t-test for the correlation coefficient

   c. p-test for the correlation coefficient

   d. r-test for the correlation coefficient

3. Which of the following results is significant at the .05 level (two-tailed test)?

   a. $r_{(30)} = .33$

   b. $r_{(60)} = .24$

   c. $r_{(4)} = .79$

   d. $r_{(10)} = .59$

4. Which of the following results is significant at the .01 level (one-tailed test)?

   a. $r_{(30)} = .42$

   b. $r_{(10)} = .62$

   c. $r_{(4)} = .87$

   d. $r_{(5)} = .81$

5. If I suspected that college education has no impact on the student's earnings five years after leaving school, which of the following would I choose?

 a. $H_0: \rho_{xy} = 0$

 b. $H_0: \rho_{xy} = 1$

 c. $H_1: r_{xy} \neq 0$

 d. $H_1: r_{xy} \neq 1$

 e. $H_1: r_x > r_y$

6. A researcher looking at secure attachments among mothers and their children finds that the more independent the children are, the less upset they are when their mother leaves the room, $r_{(50)} = -.633, p < .05$. How many participants are in this study?

 a. 25

 b. 50

 c. 52

 d. 100

7. A researcher measures defensiveness among 6-, 7-, and 8-year-olds accused of cheating on a spelling test. He finds that 8-year-olds tend to be more defensive than both 6- and 7-year-olds. What analysis did the researcher most likely run?

 a. Pearson's correlation coefficient

 b. A t-test for independent means

 c. A t-test for dependent means

 d. A simple ANOVA

SHORT-ANSWER/ESSAY QUESTIONS

1. What is the critical value of the correlation coefficient needed for rejection of the null hypothesis if your degrees of freedom is 15 and you are conducting a one-tailed test using the .01 level of significance?

2. What is the critical value of the correlation coefficient needed for rejection of the null hypothesis if your degrees of freedom is 30 and you are conducting a two-tailed test using the .05 level of significance?

3. If your total sample size is 75, what is your degrees of freedom for the correlation coefficient?

4. Your degrees of freedom is 55. Looking at the table of critical values for the correlation coefficient, you can see that there are only entries for 50 and 60 degrees of freedom, not 55. If you want to be more conservative, which entry should you choose?

5. How would you interpret $r_{(37)} = .89, p < .05$?

6. Did you know that there is a correlation between sour cream consumption and motorcycle accidents? Did you know that there is a correlation between math degrees and suicide (hopefully none of you in this math-based class!). I want you to give me 1) an example of a positive (direct) correlation and 2) a negative (indirect) correlation. Don't just make

these up. Find some evidence for them. My advice is to use the Internet. There are good examples if you use search terms "crazy" or "spurious" correlations.

7. Can the same study look at both differences between two groups as well as the relationship between two variables? Why or why not? If yes, design a study that would allow you to look at both.

8. Imagine you find the following table correlating the number of years in college and salary (two-tailed test). Interpret the results and write it up as you would see it in a journal article.

Correlations

		Years College	Income
Years College	Pearson Correlation	1	.533**
	Sig. (2-tailed)		.007
	N	24	24
Income	Pearson Correlation	.533**	1
	Sig. (2-tailed)	.007	
	N	24	24

** Correlation is significant at the 0.01 level (2-tailed).

SPSS QUESTION

1. Using SPSS, perform the eight steps of hypothesis testing to test the null hypothesis that there is no relationship between IQ and salary, using the following set of data. Also, report your result in the following form: $r_{(17)} = .55, p > .05$.

Salary ($1,000)	IQ
245	127
120	133
90	98
88	105
75	115
74	102
66	115
58	98
45	80
23	85
21	78
15	70

JUST FOR FUN/CHALLENGE YOURSELF

1. Why do the critical values for the correlation coefficient decrease as the sample size increases? Why do they increase when you have a lower value for the possibility of Type I error?

ANSWER KEY

TRUE/FALSE QUESTIONS

1. False. The correlation coefficient can be used for both one-tailed and two-tailed tests.

2. True.

3. False. You can have a significant correlation that is very weak, meaning that there is a very weak (hence, not meaningful) relationship between the two variables.

4. False. A correlation does not mean causation. There is merely a relationship between age and authoritarianism.

5. False. Large coefficients of alienation (nondetermination) indicate less overlap, not more.

6. False. Correlations range from -1 to $+1$, so a correlation of 2.34 is not possible.

MULTIPLE-CHOICE QUESTIONS

1. (c) the relationship between two variables

2. (b) t-test for the correlation coefficient

3. (d) $r_{(10)} = .59$

4. (a) $r_{(30)} = .42$

5. (c) $H_1: r_{xy} \neq 0$

6. (c) 52. The degrees of freedom for correlations is $n - 2$, or in this case $52 - 2 = 50$

7. (d) A simple ANOVA given the three independent groups (6-, 7-, and 8-year-olds)

SHORT-ANSWER/ESSAY QUESTIONS

1. The critical value is .5577.

2. The critical value is .3494.

3. The degrees of freedom $= n - 2 = 75 - 2 = 73$.

4. To be more conservative, you choose the entry for 50 degrees of freedom, which results in a higher critical value for the correlation coefficient. This in turn makes the test more conservative, with a finding of significance less likely.

5. First, r represents the test statistic that was used, 37 is the number of degrees of freedom, and .89 represents the obtained value that was calculated for the correlation coefficient. Finally,

$p < .05$ indicates the probability is less than 5% that the relationship between the two variables is due to chance alone. We can conclude that there is a significant relationship between the two variables.

6. A direct correlation implies that as one variable increases, the other increases (or as one variable decreases, the other decreases). For example, increased class attendance correlates directly with increased grades. An indirect correlation implies that as one variable decreases another increases. For example, higher self-esteem correlates with low levels of depression.

7. Yes. Some studies look at both between-group differences as well as correlations among some of the dependent variables. For example, a researcher might assign participants to view a sexual harassment film in which the victim responds in either a hostile or submissive manner. Type of response is thus the independent variable. In a subsequent survey, the participant might rate both the likelihood of harassment (1 = no harassment to 10 = a great deal of harassment) as well as the severity of the harassment (1 = not at all severe to 10 = extremely severe). The researcher could then measure whether type of response differentially impacts both likelihood of harassment and severity of harassment ratings, but the researcher can also determine if higher likelihood of harassment ratings correlate with the higher (or lower) ratings of harassment severity.

8. The results would look like the following: "The positive correlation between years in school and income five years after leaving school was statistically significant, $r_{(22)} = .53, p < .05$, two-tailed." Keep in mind that the degrees of freedom is $n - 2$, or in this case $24 - 2 = 22$.

SPSS QUESTION

1. The eight steps of hypothesis testing:

    1. State the null and research hypotheses:

        $H_0: \rho_{xy} = 0$

        $H_1: r_{xy} \neq 0$

    2. Set the level of significance: .05.

    3. Select the appropriate statistic: the correlation coefficient.

    4. Compute the test statistic value: This is done for us automatically in SPSS. The following table illustrates the SPSS output.

Correlations

		Salary_1k	**IQ**
Salary_1k	Pearson Correlation	1	0.772**
	Sig. (2-tailed)		0.003
	N	12	12
IQ	Pearson Correlation	0.772**	1
	Sig. (2-tailed)	0.003	
	N	12	12

** Correlation is significant at the 0.01 level (2-tailed).

5. Determine the value needed for rejection of the null hypothesis: Because our sample size is 12, we know that our degrees of freedom is 10. As we are conducting a two-tailed test using the .05 level of significance, our critical value is .5760. Note: This step does not necessarily need to be done, because SPSS automatically calculates the significance level for the correlation coefficient.

6. Compare the obtained value with the critical value: Our obtained value is higher than our critical value.

7. and 8. Decision time: Because our obtained value is higher than our critical value, you can reject the null hypothesis that there is no relationship between IQ and salary. We obtained a positive correlation coefficient, indicating that there is a positive association between these two variables; in other words, higher IQ is associated with higher salary.

Additionally, this result can be reported as: $r_{(10)} = .772, p < .05$.

JUST FOR FUN/CHALLENGE YOURSELF

1. Critical values for the correlation coefficient (more generally) decrease as the sample size increases because, as we have a greater number of individuals, it is easier for a relationship of a certain strength to be seen; hence, when you have a lot of data, you don't need a very strong relationship to achieve statistical significance.

    On the other hand, when your sample size is very small, the correlation coefficient must be very high for you to be able to say that the relationship is valid and not just due to chance. Additionally, a lower value for the probability of Type I error means that you have a higher degree of certainty that the relationship is valid and not due to chance. Therefore, your correlation coefficient must be that much higher to support a greater degree of certainty.

16 Predicting Who'll Win the Super Bowl

Using Linear Regression

LEARNING OBJECTIVES

- Learn about linear regression and how it can be used for prediction.

- Understand when it is appropriate to use linear regression.

- Learn how to determine the accuracy of your predictions.

- Understand when it is and when it is not appropriate to include multiple independent variables in what is called multiple regression.

SUMMARY/KEY POINTS

- Linear regression, in essence, uses correlations between variables as the basis for predicting the value of one variable based on the value of another. For example, imagine you are interested in looking at how a judge sentences defendants convicted of burglary. If you think the judge's sentencing decisions are reflective of the defendant's prior convictions, you can look at the correlation between prior convictions and prison sentences to predict how he might sentence future defendants.
  - The higher the absolute value of the correlation coefficient, the more accurate the prediction. In the case of linear regression, a perfect correlation translates into a perfect prediction. Thus, a convict with one prior conviction might get 10 months, a convict with two priors gets 11 months, a convict with three priors gets 12 months, and so forth (1 additional month for each additional prior conviction). Here, the correlation would be a perfect 1.00. This would indicate that prior convictions alone are used in sentencing determinations.
  - In linear regression, a regression equation is determined, and this equation can be used to plot a regression line. The regression line reflects the best estimate of predicted scores for the dependent variable based on levels of the independent variable. A perfect correlation would result in a regression line with a 45° angle. That is, for each prior conviction, convicts are sentenced to one additional month in prison.
    - In the regression equation, the predicted score of the dependent variable is equal to the slope multiplied by the value of the independent variable, plus a constant that is equal to the point at which the regression line crosses the y-axis.
  - Error in prediction (error of estimate) is calculated as the distance between each individual data point and the regression line—the error value indicates by how much your prediction was "off." The regression line is thus drawn to minimize the distance between the line itself and each score in the scatterplot. If there is a perfect correlation, then each data point will fall "on the line," though this is very rare, especially in psychology (two prior convictions might be seen as much worse than just one prior, with the latter getting a much higher sentence. Yet a judge may see ten and eleven priors as similarly bad, and he may sentence both convicts to the same sentence although one has more priors than the other).
    - Standard error of estimate is the average of all values for error in prediction. This value tells you how imprecise the predictive power of the linear regression analysis is overall.
    - Error decreases as the correlation between the two variables increases. The closer to 1.00 (or the closer to −1.00 for a negative correlation), the less error in the design. Perfect correlations, of course, are very rare in psychology, but you might find them in other areas. The Celsius and Fahrenheit temperature scales, for example, correlate perfectly, but that is primarily because they measure the same thing (temperature!)

- In linear regression, the outcome variable is called the criterion or dependent variable and written as Y (sentence in our sentencing example), while the predictor or independent variable is written as X (prior convictions).

- In multiple regression, more than one independent variable is included in the analysis.
  - An additional independent variable should only be included if it makes a unique contribution to the understanding, or prediction, of the dependent variable. In our sentencing example, maybe we know that the judge is known to be "moody." We can look at judge moodiness as a factor in sentencing. We can similarly consider the viciousness of the crime, the age of the defendant, the dollar value of burgled goods, etc.
  - Additionally, multiple independent variables are best included in a regression analysis if they are uncorrelated with one another but are all correlated with the dependent variable.

KEY TERMS

- **Regression equation**: In regression, an equation that defines the line that has the best fit with your data

- **Regression line**: The line of best fit that is drawn (or calculated) based on the values in the regression equation

- **Line of best fit**: The regression line that best fits the data and minimizes the error in prediction

- **Error in prediction**: (aka error of estimate): The difference between the actual score and the predicted score in a regression

- **Criterion or dependent variable**: The outcome variable, or the variable that is predicted, in a regression analysis

- **Predictor or independent variable**: The variable that is used to predict the dependent variable in a regression analysis

- **Y prime**: The predicted value of Y, the dependent variable, written as Y'

- **Standard error of estimate**: The average amount that each data point differs from the predicted data point

- **Multiple regression**: A type of regression in which more than one independent variable is included in the analysis

TRUE/FALSE QUESTIONS

1. Linear regression uses correlations as its basis.

2. Linear regression can be used to predict values of the dependent variable for individuals outside of your data set.

3. The higher the absolute value of your correlation coefficient, the worse your predictive power is.

4. When using multiple regression, it is always best to include as many predictor variables as possible.

5. When using multiple regression, it is best to select independent variables that are uncorrelated with one another but are all related to the predicted variable.

6. A researcher studying how older adults cope with their own death predicts that if older adults look back and see a successful life, they will fear death less. If they look back and find success lacking in their life, they will fear death more. The predictor variable in this study is fear of death.

7. A therapist wants her clients to write their thoughts and feelings in a journal throughout the day. In the past, she has found that the more journal entries a client writes, the better able they are at coping with life frustrations (0 = lowest coping level to 10 = highest coping level). She wants to make sure a current client has a coping score of 7 by the end of the week, so she tells the client to write exactly ten journal entries each day (the number of entries that aligns with a coping score of 7 in a regression equation). If the client follows her advice perfectly, the client is guaranteed to have an exact score of 7 by the end of the week.

8. A researcher thinks that a child's locus of control score will predict whether he asks for help from a teacher. In the formula $Y' = bX + a$, the letter X represents the locus of control score.

MULTIPLE-CHOICE QUESTIONS

1. Your prediction in linear regression will be perfect if your correlation is _____.

   a. -1

   b. $+1$

   c. 0

   d. either a or b

2. If the prediction in a linear regression analysis were perfect, all predicted points would fall _____.

   a. on the regression line

   b. above the regression line

   c. below the regression line

   d. both above and below the regression line

3. If your standard error of estimate is high, the plot of the regression line with data points will show _____.

   a. data points very close to the regression line

   b. data points very far from the regression line

   c. data points exactly on the regression line

   d. any of a, b, or c—it cannot be determined

4. The variable that is predicted in linear regression is called _____.

    a. the criterion variable

    b. the dependent variable

    c. the predictor variable

    d. the independent variable

    e. both a and b

5. The variable that is used to predict another variable in linear regression is called the _____.

    a. criterion variable

    b. dependent variable

    c. predicted variable

    d. independent variable

6. In linear regression, the dependent variable is indicated by which of the following?

    a. Y

    b. b

    c. X

    d. a

7. In linear regression, the independent variable is indicated by which of the following?

    a. Y

    b. b

    c. X

    d. a

8. In linear regression, the slope is indicated by which of the following?

    a. Y

    b. b

    c. X

    d. a

9. In linear regression, the point at which the line crosses the y-axis is indicated by which of the following?

    a. Y

    b. b

    c. X

    d. a

10. The predicted value of Y is called _____.

    a. Y prime

    b. Y sigma

    c. Y delta

    d. Y alpha

11. In linear regression, the regression line can be _____.

    a. a straight line only

    b. a straight or curved line

    c. a curved line only

    d. none of the above

12. If you have a correlation coefficient of -1, the standard error of estimate will be equal to which of the following?

    a. 1

    b. -1

    c. 2

    d. 0

13. If you found a perfect positive correlation between the hours per week you spent studying and your final grade in this course, what should the trend line look like?

    a. It should slant upward from left to right (start low and get higher at a 45-degree angle).

    b. It should slant downward from left to right (start high and get lower at a 45-degree angle).

    c. It should stay flat at a horizontal angle.

    d. It should be curvilinear (start low, get high, and end low, or start high, get low, and end high).

14. In the regression equation $Y' = bX + a$, what is the slope?

    a. a

    b. b

    c. X

    d. Y'

15. A researcher is interested in seeing what factors correlate with alcohol abuse. He knows loneliness is one contributing factor, so he includes that in a multiple regression model. He is also considering the following factors, but which one should he avoid:

    a. Work stress

    b. Coping ability

    c. Social isolation

    d. Peer pressure

16. Jeff has an above average intelligence score as calculated by the Stanford–Binet Intelligence Scales. If there is a strong, positive correlation between the Stanford–Binet Intelligence Scales and the Wechsler intelligence test, predict how he would perform on the Wechsler test.

    a. Excellent

    b. Above average

    c. Average

    d. Below average

EXERCISES

1. Using the following data, conduct a linear regression analysis (by hand). In this analysis, you are testing whether SAT scores are predictive of overall college GPA. Also, write out the regression equation. How would you interpret your results?

SAT Score	GPA
670	1.2
720	1.8
750	2.3
845	1.9
960	3.0
1,000	3.3
1,180	3.2
1,200	3.4
1,370	2.9
1,450	3.8
1,580	4.0
1,600	3.9

2. You conduct a linear regression in which the number of hours per week spent exercising is used to predict the respondent's overall general health, measured on a scale from 0 to 100 with 100 being the best possible health. You obtain the following results: $b = 4.5$; $a = 37$. Using the general formula for a regression line, calculate the predicted values for overall general health for respondents who spend 0 hours per week exercising, 3 hours per week exercising, and 12 hours per week exercising.

SHORT-ANSWER/ESSAY QUESTIONS

1. When you are deciding which variables to include as predictors in a multiple regression equation, what are some conditions that you must consider first?

2. Describe how correlation and regression are linked yet distinct.

3. A sleep therapist finds that prescribing high doses of benzodiazepine to patients suffering from chronic insomnia lowers their rates of sleeplessness by exactly 78.54%. Is this specific prediction warranted? Why or why not?

4. In *Roper vs Simmons*, the US Supreme Court ruled that juvenile defendants can no longer be sentenced to death in a capital trial. They cited information submitted by the American Psychological Association showing that age is correlated with impulsiveness (young people are often impulsive, a factor that might lead them to commit a homicide). Imagine the Court is unsure whether to categorize 16-, 17-, 18-, or 19-year-olds as adults or as juveniles. However, they do have an impulsiveness threshold. If the defendant's impulsiveness score (Y') is lower than or equal to an impulsiveness threshold of 50 set by the Court, the defendant will be considered a juvenile and thus will be ineligible for the death penalty. Using X for age (the predictor variable) with a slope (b) of 2 an α of 14, are 19-year-olds eligible for the death penalty? What about 16-, 17-, and 18-year-olds?

SPSS QUESTION

1. Using the data presented under the "Exercises" section, question 1, conduct the same linear regression using SPSS and interpret the output. Additionally, create a scatterplot of the data with a superimposed regression line.

JUST FOR FUN/CHALLENGE YOURSELF

1. Starting with the data from the "Exercises" section, question 2, you find two new individuals who have health scores of 100 and 75, respectively. Calculate the predicted values for the number of hours they exercise per week.

2. Using the following data, conduct a multiple linear regression in SPSS and interpret the results. SAT score and IQ are both independent variables, and GRE score is the dependent variable. Interpret the results and write out the regression equation.

SAT Score	IQ	GPA
670	80	1.2
720	87	1.8
750	105	2.3
845	95	1.9
960	110	3.0
1,000	98	3.3
1,180	110	3.2
1,200	125	3.4
1,370	115	2.9
1,450	120	3.8
1,580	140	4.0
1,600	135	3.9

ANSWER KEY

TRUE/FALSE QUESTIONS

1. True.

2. True.

3. False. The higher the absolute value of your correlation coefficient, the better your predictive power is.

4. False. Careful judgment should be used when deciding which predictor variables to include.

5. True.

6. False. The predictor (independent) variable is perceived life success. Fear of death is the criterion (dependent) variable.

7. False. Regression is used to predict scores, but unless there is a perfect correlation (which is very rare), a guaranteed prediction is never warranted. The client's score will most likely vary from the predicted score, dependent on the standard error of estimate.

8. True.

MULTIPLE-CHOICE QUESTIONS

1. (d) either a or b

2. (a) on the regression line

3. (b) data points very far from the regression line

4. (e) both a and b

5. (d) independent variable

6. (a) Y

7. (c) X

8. (b) b

9. (d) a

10. (a) Y prime

11. (a) a straight line only

12. (d) 0

13. (a) It should slant upward from left to right (start low and get higher at a 45-degree angle).

14. (b) b

15. (c) Social isolation. Since this overlaps with loneliness, it may contribute very little as a predictor.

16. (b) Above average

EXERCISES

1. The regression coefficients are calculated using the following equations:

$$b = \frac{\sum XY - \left(\sum X \sum y / n\right)}{\sum X^2 - \left[\left(\sum X\right)^2 / n\right]} \Rightarrow$$

$$\frac{41,509.5 - (13,325 \times 34.7 / 12)}{16,033,625 - 13,325^2 / 12}$$

$$a = \frac{\sum Y - b \sum X}{n} \Rightarrow$$

$$\frac{34.7 - 0.002 \times 13,325}{12} = 0.219$$

The regression equation would be represented as follows:

$$Y' = 0.002X + 0.219$$

The value for a, 0.219, indicates the predicted GPA if the SAT score were equal to zero. The value for b, 0.002, indicates that a 1-unit increase in SAT score is associated with a 0.002 predicted increase in GPA. Multiplying this figure by 100, we can say that a 100-unit increase in SAT score is associated with a 0.2 predicted increase in GPA. This means, more generally, that higher SAT scores are associated with higher GPAs.

2. Here is the regression equation:

$$Y' = 4.5X + 37$$

For respondents who spend 0 hours per week exercising:

$$Y' = 4.5(0) + 37 = 37$$

For respondents who spend 3 hours per week exercising:

$$Y' = 4.5(3) + 37 = 40.5$$

For respondents who spend 12 hours per week exercising:

$$Y' = 4.5(12) + 37 = 81$$

SHORT-ANSWER/ESSAY QUESTIONS

1. In deciding which predictor variables to add into a regression equation, keep in mind a number of considerations. You must think about whether a new variable will make a unique contribution to understanding the dependent variable. The two (or more) variables in combination should predict Y better than any of the variables do alone. You must also balance the cost of resources needed for another predictor with the possible benefit of adding another predictive value. Additionally, there must be a theoretical (or literature-based) rationale for adding another predictor; there is a limit to how many variables will theoretically contribute to a prediction. Finally, when including multiple independent/predictor variables in a regression analysis, it is best if they are uncorrelated with one another but are both correlated to the dependent variable.

2. Correlation is used to report only the relationship between two variables, whereas in regression, one variable is designated as the independent/predictor variable and the other variable as the dependent/criterion variable. This designation allows regression to be used to predict the value of one variable from the value of another. Correlation is used in the steps of conducting a regression. In fact, linear regression uses correlations between variables as the basis for predicting the value of one variable based on the value of another. In linear regression, a perfect correlation would translate into perfect prediction. In multiple regression, correlations are run as a preliminary step to assure the investigator that the independent/predictor variables are not strongly correlated with one another and to determine that each independent/predictor variable is correlated with the dependent/criterion variable. Finally, neither correlation nor regression results can be used to claim causality between variables.

3. A precise prediction is not warranted, as the prediction is merely an estimate. The true effectiveness of the dosage will hopefully fall along the regression trend line, but this is not guaranteed.

4. Recall the regression formula: $Y' = bX + \alpha$. Plugging in values to $Y' = bX + \alpha$, where the maturity threshold is 50, Y' is unknown, $b = 2$, and $\alpha = 14$, we get $2X + 14$. If we replace X with 19, then $2(19) + 14 = 52$. A 19-year-old's score of 52 is predicted to be above the maturity threshold, and thus the 18-year-old will be eligible for the death penalty. If we replace X with 18, then $2(18) + 14 = 50$, making the 18-year-old right at the maturity threshold of 50. Since the defendant needs to be higher than the threshold, he or she would not be death eligible. The same would apply to 17-year-olds (maturity score at 48, under the 50 threshold) and 16-year-olds (maturity at 46, under the 50 threshold).

SPSS QUESTION

1. The SPSS output is shown here:

Model Summary

Model	R	R Square	Adjusted R Square	Std. Error of the Estimate
dimension0 1	.893[a]	.797	.777	.42679

[a] Predictors: (Constant), SAT

ANOVA[b]

Model	Sum of Squares	df	Mean Square	F	Sig.
Regression	7.168	1	7.168	39.351	.000[a]
	1.821	10	.182		
Residual Total	8.989	11			

[a] Predictors: (Constant), SAT
[b] Dependent Variable: GRE

Coefficients[a]

Model		Unstandardized Coefficients		Standardized Coefficients		
		B	Std. Error	Beta	t	Sig.
1	(Constant)	.219	.444		.494	.632
	SAT	.002	.000	.893	6.273	.000

[a] Dependent Variable: GPA

The final table, "Coefficients," gives us our *a* and *b* values.

The value for *a*, 0.219, indicates the predicted GPA if the SAT score were equal to zero. The value for *b*, 0.002, indicates that a 1-unit increase in SAT score is associated with a 0.002 predicted increase in GPA. Multiplying this figure by 100, we can say that a 100-unit increase in SAT score is associated with a 0.2 predicted increase in GPA. This means, more generally, that higher SAT scores are associated with higher GPAs. Additionally, the effect of SAT was found to be significant at the .05 level of significance.

The scatterplot with superimposed regression line is shown here:

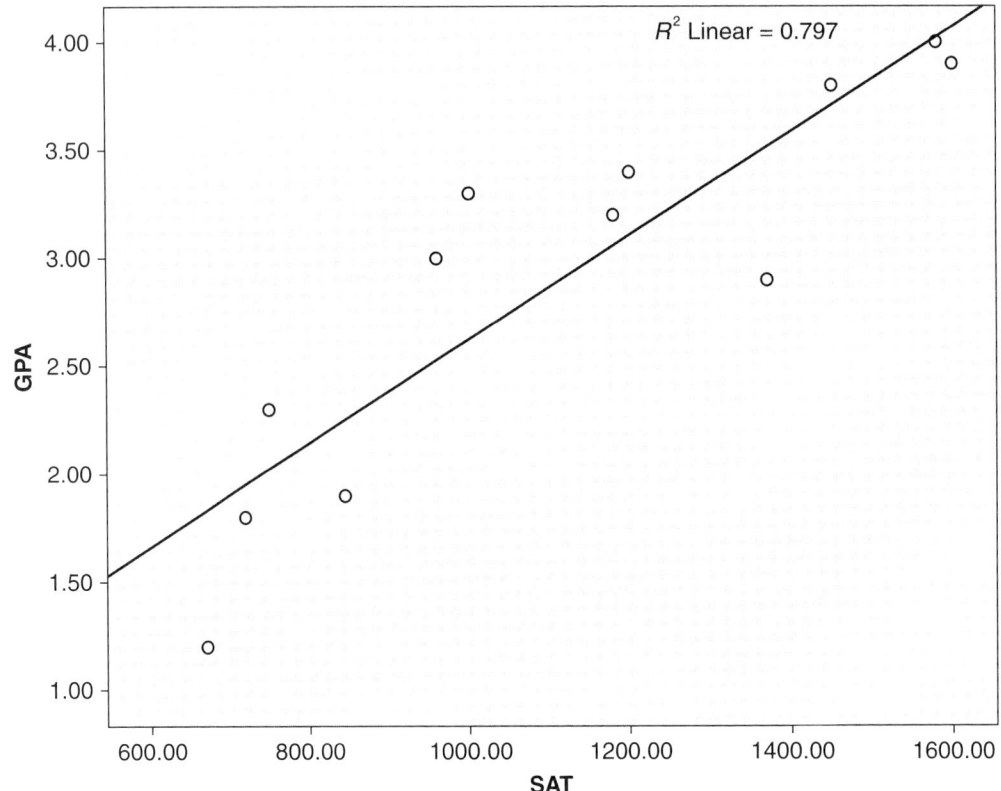

JUST FOR FUN/CHALLENGE YOURSELF

1. The results were $b = 4.5$ and $a = 37$. These values give us this equation:

$$Y' = 4.5X + 37$$

For the respondent with a health score of 100,

$$Y' = 4.5X + 37 \Rightarrow$$
$$100 = 4.5X + 37$$
$$63 = 4.5X$$
$$X = 14$$

For the respondent with a health score of 75,

$$Y' = 4.5X + 37 \Rightarrow$$
$$75 = 4.5X + 37$$
$$38 = 4.5X$$
$$X = 8.44$$

2. The SPSS output is shown here:

Model Summary

Model	R	R Square	Adjusted R Square	Std. Error of the Estimate
1	.917[a]	.841	.805	.39901

[a] Predictors: (Constant), IQ, SAT

ANOVA[b]

Model	Sum of Squares	df	Mean Square	F	Sig.
Regression	7.556	2	3.778	23.731	.000[a]
	1.433	9	.159		
Residual Total	8.989	11			

[a] Predictors: (Constant), IQ, SAT
[b] Dependent Variable: GPA

Coefficients[a]

Model		Unstandardized Coefficients		Standardized Coefficients		
		B	Std. Error	Beta	t	Sig.
11	(Constant)	−1.144	.966		−1.184	.267
	SAT	.001	.001	.432	1.334	.215
	IQ	.025	.016	.506	1.562	.153

[a] Dependent Variable: GPA

Neither SAT nor IQ were found to be significant. However, based on the coefficients, a 1-unit increase in SAT is associated with a 0.001-unit increase in GPA, while a 1-unit increase in IQ is associated with a 0.025-unit increase in GPA. The value for a, −1.144, indicates the predicted GPA if the SAT score and IQ were equal to zero.

Here is the regression equation: $Y' = 0.001(SAT) + 0.025(IQ) − 1.144$

17 What to Do When You're Not Normal

Chi-Square and Some Other Nonparametric Tests

LEARNING OBJECTIVES

- Understand the reason behind the use of nonparametric statistics and when they should be preferred.

- Learn about chi-square and how it is calculated, both by hand and by using SPSS.

- Briefly review some other nonparametric statistics.

SUMMARY/KEY POINTS

- The statistical tests covered previously in this book consisted of parametric statistical tests, which make certain assumptions about the data (which should be homogenous), sample size (which should be large), the type of scale used (which should be interval or ratio), and normality (which indicates a normal curve). If you wanted to see if elderly adults exposed to pets find life more meaningful than elderly adults not exposed to pets, you could see if the two levels of the independent variable (pet exposure) differentially impact the elderly adults' ratings of life meaningfulness (the dependent variable).

- Nonparametric statistics do not incorporate these same assumptions. These types of statistics can be used when the assumptions of parametric statistical tests have been violated.
  - Nonparametric statistics can be preferred when you have sample sizes that are very small (less than 30) or when you are analyzing categorical variables. For example, if you ask the simple question, "Is your life meaningful" with a yes or no (nominal) dependent variable, you cannot run a *t*-test or ANOVA. After all, you cannot calculate the mean using a categorical yes/no question, and the *t*-test and ANOVA require the mean. However, you can use a nonparametric test, which focuses on frequencies rather than a comparison of means.

- The nonparametric test highlighted in this chapter is chi-square.
  - A one-sample chi-square focuses on the distribution of a single variable. It is used to determine whether the distribution of a single categorical variable is significantly different from what would be expected by chance. For example, imagine a police department is assessing traffic stops by police officers. They want to see if the expected stops align with the observed stops in a diverse community. The researchers would then compare the number of observed stops of Caucasian, African American, and Other groups against the expected number of stops. If they differ, the researcher might conclude that racial discrimination may be present.
  - The one-sample chi-square test is also called goodness of fit. That is, does the percentage of African Americans stopped *fit* the percentage of African Americans in the community?
  - A two-sample chi-square focuses on the relationship between two categorical variables. It determines whether the variables are significantly related, or dependent. For example, a researcher might determine the extent to which moviegoers who just watched either a cartoon or a violent movie will help (or not help) someone with crutches who has fallen down outside the movie theater.
  - The two-sample chi-square test is also called the test of independence. That is, is the type of movie a person watches *independent* of whether they help?
  - The chi-square statistic measures the difference between the observed data and what would be expected by chance alone.

- The chapter also lists nine other nonparametric tests, when the tests should be used, and sample research questions for each.

KEY TERMS

- **Parametric statistics**: A set of statistical tests that incorporate certain assumptions, often including, among others, a sufficient sample size and a normal or near-normal distribution of continuous variables

- **Nonparametric statistics** (aka distribution-free statistics): A set of statistical tests that do not incorporate the assumptions held by parametric statistical tests

- **Chi-square**: A nonparametric statistical test used to determine whether a single variable has a distribution that would be expected through chance or whether a general relationship exists between two categorical variables.

TRUE/FALSE QUESTIONS

1. Nonparametric statistics include more assumptions than do parametric statistics.

2. Chi-square is a nonparametric statistical test.

3. A chi-square (rather than a *t*-test) should be used to test the question, "Are employers more likely to hire or not hire an attractive applicant than an unattractive applicant?"

4. A researcher is interested in evaluating patient discharge decisions by insight therapists with different specialization backgrounds. He gives therapists with a psychodynamic background, a cognitive background, or a humanistic background patient files and asks whether the patients should or should not be discharged. The researcher should analyze the data with a chi-square test of independence.

5. A researcher studying satiation (the feeling of "fullness" after a meal) thinks that people feel more full (1 = not at all full to 9 = very full) after finishing a 600-calorie meal placed on a small diameter plate compared to a 600-calorie meal placed on a large diameter plate. The researcher should analyze the data with a chi-square test of independence.

MULTIPLE-CHOICE QUESTIONS

1. Which of the following tests is also known as goodness of fit?

    a. One-sample chi-square

    b. Two-sample chi-square

    c. Any nonparametric statistical test

    d. Fisher's exact test

2. Which of the following tests is used to determine whether the distribution of a single categorical variable significantly differs from what would be expected by chance?

    a. Two-sample chi-square

    b. Fisher's exact test

    c. Kolmogorov-Smirnov test

    d. One-sample chi-square

3. Which of the following tests is used to determine whether two categorical variables are related, or dependent?

    a. Two-sample chi-square

    b. The sign test

    c. Kolmogorov-Smirnov test

    d. One-sample chi-square

4. Nonparametric statistics might be preferred under which of the following conditions?

    a. Your sample size is very small.

    b. Your sample size is very large.

    c. The variables you are analyzing are continuous.

    d. All the assumptions of parametric statistics have been met.

5. If there is no difference between what is observed and what is expected, your chi-square value will be _____.

    a. 1

    b. 0

    c. −1

    d. a, b, or c—it is impossible to determine

6. Which of the following values is significant at the .05 level?

    a. $\chi^2_{(10)} = 15.43$.

    b. $\chi^2_{(8)} = 12.20$.

    c. $\chi^2_{(5)} = 11.12$.

    d. $\chi^2_{(2)} = 4.23$.

7. Which of the following values is significant at the .01 level?

    a. $\chi^2_{(12)} = 22.14$.

    b. $\chi^2_{(6)} = 14.20$.

    c. $\chi^2_{(4)} = 8.42$.

    d. $\chi^2_{(2)} = 9.45$.

8. In a study on stereotype threats, a researcher predicts that women will get a lower percentage of math questions correct when men are present than when only women are present. What test should the researcher use to analyze this study?

    a. Chi-square test for independence

    b. Chi-square goodness of fit test

    c. t-test

    d. Simple ANOVA

9. In a bait-and-switch study, researchers tell participants expecting to participate in an interesting study for $5 that the study has been cancelled. When asked if they would

participate in a less-interesting study for no money, they find that 47% of the sample agrees to participate (compared to 15% who never knew about the original study). If the researcher finds $X^2_{(1)} = 7.13$, what should the researcher conclude?

a. There is no significant difference between those who knew about the original study and those who did not, $p > .05$.

b. There is a significant difference between those who knew about the original study and those who did not, $p < .05$.

c. There is a significant difference between those who knew about the original study and those who did not, $p < .01$.

d. The researchers should not have analyzed this data with a chi-square.

10. You are taking a multiple-choice exam and find that there are a lot of answers with A, B, and D as the correct option, but very few with C as the correct option. You start to think that you probably got some prior questions wrong, because some of them have to be C, right! This assumes, of course, that your instructor wants to have an equivalent number of answers related to options A, B, C, and D. After the exam, you run this idea by the instructor, who is also curious about whether he neglects the option C. What test should you run to see if the answer options are equivalent?

a. Chi-square test for independence

b. Chi-square goodness of fit test

c. *t*-test

d. Simple ANOVA

11. A researcher expects half of his 100 participant sample to be male and half to be female. He finds 60 males and 40 females. Did he support the null hypothesis?

a. Yes, the null hypothesis should be retained, $p > .05$.

b. No, the null hypothesis should not be retained, $p < .05$.

c. No, the null hypothesis should not be retained, $p < .01$.

d. There is not enough information here to decide whether the null hypothesis should be retained.

12. A researcher recruits those suffering from various phobias to participate in a study. She has those suffering from arachnophobia (fear of spiders), cynophobia (fear of dogs), dentrophobia (fear of dentists), glossophobia (fear of public speaking), and phobophobia (fear of phobias). She hopes to have equal numbers of participants representing each phobia in the study, and finds no difference in the chi-square ($p > .05$). What would her degrees of freedom be in this study?

a. 1

b. 2

c. 3

d. 4

e. 5

EXERCISE

1. Using the following data, perform the eight steps of hypothesis testing. In this question, calculate the chi-square statistic by hand.

Expected Number	Observed Number
10	15
20	15
20	37
50	33

SHORT-ANSWER/ESSAY QUESTION

1. Interpret the following: $\chi^2_{(3)} = 14.2, p < .05$.

2. Describe a study in which you can use both nonparametric and parametric statistics.

3. A researcher has participants expecting to receive either painful or mild electrical shocks during a study wait in a quarter-full waiting room with other participants. He then measures whether they choose to sit next to other participants or if they sit far away on their own. What is the best test to run for this study and why?

SPSS QUESTION

1. Using the following data, perform a chi-square analysis using SPSS. Interpret your results.

Expected Number	Observed Number
10	8
10	10
10	11
10	9

JUST FOR FUN/CHALLENGE YOURSELF

1. What test can be used to see whether scores from a sample come from a specified population?

2. Which test computes the exact probability of outcomes in a 2 × 2 table?

3. Which test computes the correlation between ranks?

TRUE/FALSE QUESTIONS

1. False. Nonparametric statistics include fewer assumptions than do parametric statistics.

2. True.

3. True. The categorical dependent variable is hire versus not hire, so a chi-square is more appropriate than a *t*-test.

4. True. The chi-square test of independence measures two different dimensions at the nominal level to see if they are related. Here, the researcher wants to determine whether discharge decisions are related to therapist specialty.

5. False. The dependent variable is fullness on a 1-to-9 scale, and therefore a *t*-test is more appropriate.

MULTIPLE-CHOICE QUESTIONS

1. (a) One-sample chi-square

2. (d) One-sample chi-square

3. (a) Two-sample chi-square

4. (a) Your sample size is very small.

5. (b) 0

6. (c) $\chi^2_{(5)} = 11.12$.

7. (d) $\chi^2_{(2)} = 9.45$.

8. (c) *t*-test. The dependent variable is measured on a ratio scale (0% correct to 100% correct).

9. (c) There is a significant difference between those who knew about the original study and those who did not, $p < .01$.

10. (b) Chi-square goodness of fit test

11. (b) No, the null hypothesis should not be retained, $p < .05$ only. Here, $X^2_{(1)} = 4.00$, which exceeds the $p < .05$ critical value (3.84) but not the $p < .01$ critical value (6.64).

12. (d) 4, or $r - 1$ (where r = the number of rows), or $5 - 1 = 4$

EXERCISE

1. The eight steps of hypothesis testing:

    1. State the null and research hypotheses:

    $$H_0: P_1 = P_2 = P_3 = P_4$$
    $$H_1: P_1 \neq P_2 \neq P_3 \neq P_4$$

    2. Set the level of significance: 0.05.

    3. Select the appropriate test statistic: chi-square.

    4. Compute the test statistic value:

Expected Number	Observed Number
10	15
20	15
20	37
50	33

$$\chi^2 = \sum \frac{(O-E)^2}{E} \Rightarrow$$

$$\frac{(15-10)^2}{10} + \frac{(15-20)^2}{20} + \frac{(37-20)^2}{20} + \frac{(33-50)^2}{50} = 23.98$$

5. Determine the value needed to reject the null hypothesis: Critical $\chi^2_{(3)} = 7.82$.

6. Compare the obtained value with the critical value: The obtained value, 23.98, is larger than the critical value of 7.82.

7. and 8. Decision time: Because the obtained value is greater than the critical value, we reject the null hypothesis that there is no difference between these groups.

SHORT-ANSWER/ESSAY QUESTION

1. First, χ^2 (chi-square) represents the test statistic. The value of 3 represents the degrees of freedom, while 14.2 is the obtained value arrived at by using the formula for chi-square. Finally, $p < .05$ indicates that the probability is less than 5% that the variable is equally distributed across all categories.

2. In experimental research, the researcher often manipulates the independent variable to determine whether it affects scores on the dependent variable. They can run inferential statistics to test this relationship. Yet sometimes they need to make sure the manipulation worked as intended. For example, in a study looking at the impact of gory crime scene pictures on jurors, I might warn some participants about the picture and not warn others. Even though my goal might be to see how the picture impacts their verdict confidence (1 = I am more confident in guilt to 9 = I am less confident in guilt), I might also want to see if they recall being warned or not. I would use parametric statistics to evaluate the verdict confidence variable and nonparametric statistics to evaluate the recall of the warning.

3. The researcher should use a chi-square test of independence since both variables are nominal (painful versus mild shock and wait alone or wait close together). Results from a similar study found that participants expecting a painful shock sat closer to other participants, perhaps reinforcing the notion that "misery loves company!"

SPSS QUESTION

1. The SPSS output is shown in the following table. The results indicate that this test did not find statistical significance, meaning that the distribution of the variable cannot be said to differ significantly from that expected through chance alone.

Test Statistics

	Var1
Chi-square	.600[a]
Df	3
Asymp. Sig.	.896

[a] *Zero cells (0.0%) have expected frequencies less than 5. The minimum expected cell frequency is 10.0.*

JUST FOR FUN/CHALLENGE YOURSELF

1. The Kolmogorov-Smirnov test

2. Fisher's exact test

3. The Spearman rank correlation coefficient

18

Some Other (Important) Statistical Procedures You Should Know About

- Review some more advanced statistical procedures and when and how they are used.

SUMMARY/KEY POINTS

- This chapter presents an overview of some more advanced statistical tests.

- MANOVA (multiple analysis of variance) is used when you want to include more than one dependent variable in an analysis. This can be preferred when the dependent variables are related to one another, because it is difficult to determine the effect of the treatment or independent variable on any one outcome. One study, for example, used student status as an independent variable (US versus international graduate students), and looked at four dependent variable scores related to achievement motivation, life satisfaction, academic stress, and locus of control, finding that US students had higher achievement motivation scores than international students.

- Repeated measures analysis of variance is used when participants are tested more than once on one factor. This might involve one independent variable with three or more levels. For example, a researcher may test children's language skills at the beginning of an academic year, halfway through the academic year, and at the end of the academic year. Researchers can also mix designs, or have one or more independent variables that are based on repeated-measures means as well as one or more independent variables based on independent means. For example, in a 2 X 2 mixed design, researchers can assign some participants to take a walk in the park while other, completely different participants take a walk downtown (independent means). For the repeated-measures means, the researchers can measure the mood of both sets of participants before and after the walk.

- Analysis of covariance is used to control for the effect of one or more continuous variables. Consider the guilt that parents of young children often experience when work interferes with family life. Researchers may predict that mothers experience more guilt than fathers, but the researchers may want to control for general levels of everyday life guilt so they can better concentrate on work-interfering-with-family guilt. Using "general" guilt as a covariate, the researchers can then see whether guilt specific to work and family life is different between mothers and fathers. (It is! Mothers experience more guilt than fathers, even when controlling for general life guilt.)

- Multiple regression is a type of linear regression that includes more than one independent variable. For example, are you addicted to games on your smartphone? Researchers might be interested in looking at factors like narcissism, anxiety, sensation seeking, neuroticism, and parental relationship (among others) to see which, if any, lead to problematic smartphone use among children.

- Meta-analysis involves combining data from several studies so you can examine patterns and trends. Think of this as a "study of studies." Looking at hundreds of sexual harassment studies, one meta-analysis found that women tend to find more evidence of sexual harassment than men. Another stress-based meta-analysis focusing on 26 studies that induced stress in a laboratory setting found that such stress leads to poor decision-making, more reward-seeking behaviors, and more risk-taking behaviors among participants.

- Discriminant analysis is used to see what sets of variables discriminate between two sets of individuals. Are you a dog person or a cat person? Discriminant analysis research shows some differences among cat and dog owners, with dog owners taking their pets for rides in the car or on visits to other people more often than cat owners. Yet both cat and dog owners felt similar emotional attachments to their pets, and they perceived similar costs of pet ownership (cleaning up after the pet, giving the pet medical attention, etc.).

- Factor analysis is used to determine the closeness of a related a set of items and determine how the items might form clusters or factors. These factor "clusters" tend to be better at representing outcomes than individual variables. For example, researchers using a confirmatory factor analysis found that five of six well known "positive" mental health scales focusing on social support, happiness, life satisfaction, resilience, and positive mental health formed a robust cluster across multiple countries (a sixth scale, focusing on optimism, did not factor into the cluster).

- Path analysis studies causality between variables. One path analysis study assessed the impact of extraversion and applicant gender on job interviewer assessments. One "path" showed that extraversion was directly correlated with a firm handshake, with this firm handshake then directly correlated with positive interview assessments. Being "male" was also directly correlated with a firm handshake, which again was directly correlated with a positive interview assessment, though gender in and of itself was not directly associated with the interview assessment. Such an analysis thus shows that a handshake mediates the relationship among predictor variables like gender and extraversion and interviewer assessments.

- Structural equation modeling (SEM) is an extension of path analysis and is also considered a generalization of regression and factor analysis. SEM is considered confirmatory as opposed to exploratory. Research using SEM looking at attitudes toward the homeless, for example, found more quality contact with the homeless led to less anxiety and more positive attitudes about the homeless. The quantity of contact, however, did not impact evaluations of the homeless.

19 Data Mining

An Introduction to Getting the Most Out of Your BIG Data

LEARNING OBJECTIVES

- Understand what data mining is.
- Understand how data mining is used to make sense out of very big data sets.
- Understand how to use SPSS to apply basic data mining tools.
- Apply pivot tables to the analysis of large data sets.

SUMMARY/KEY POINTS

- This chapter covers the analysis of big data, such as health care records, social media interactions, customers' purchasing patterns, and daily physical activity. For example, researchers might look at the factors that predict the severity of bicycle crashes. One such data mining study found that factors like the type of road (dirt or paved), age and gender of the cyclist, road signage, time of day, and time of year helped predict bicycle accidents.
  ○ Data mining is frequently used in sales and marketing, and is also called analytics.
  ○ When you consider using a big data set, keep in mind the quality of the data just as you would with other data sets. Data might be meticulously sought out with an eye for detail and quality or it might come from biased or incomplete sources. In psychology, the researcher needs to keep in mind that quantity does not imply quality!
  ○ Big data is a very large collection of either cases, variables, or both.
  ○ Big data sets can be pulled from the Internet, which means that it is always changing as new data becomes available. The ever-changing nature of the data thus implies that answering the same question at different times may result in different outcomes. For example, how will the bicycle crash data change once self-driving cars become more readily available on the road? Future data sets will be able to adapt to this emerging technology.

- You can do basic counts in SPSS of big data with frequencies. The path begins with Analyze—Descriptive Statistics—Frequencies, and can be done with numeric variables or text/string variables. You could look at the number of men versus women cyclists involved in accidents, or determine the number of accidents that occur on the open highway between cities, the urban streets within cities, or on the rural streets in out of the way small towns.

- You can also do basic counts in SPSS of big data with crosstabs to summarize categorical data. The path begins with Analyze—Descriptive Statistics—Crosstabs. Here, you can see if men (compared to women) are more likely to crash on urban or suburban streets. Or you could look at fatalities with regard to gender and whether the cyclist was wearing a helmet.

- You can use pivot tables—with frequencies, crosstabs, or other output—in SPSS to further mine your data.

KEY TERMS

- **Exabytes**: About 1 quintillion bytes; introduced in this chapter to alert readers to how very large the possibilities are for data mining

- **Data mining**: Looking for patterns in large data sets; also called analytics

- **Cross-tabulation tables**: Known as crosstabs, summarize categorical data to create column and row totals and cell counts

- **Pivot table**: Table that allows you to easily visualize and manipulate rows and columns, as well as the cells' contents

TRUE/FALSE QUESTIONS

1. Data mining is used by online stores to recommend future purchases based on your current purchases.

2. When using big data sets from Internet sources, expect the data within to be stable from month to month.

3. Procedures for mining big data can only be used for big data sets.

4. A psychologist who wants to look at how many migrants settle in a new country over the course of a two-year period should use a crosstabs procedure.

5. A psychologist can data mine to determine whether the number of homicides that occur in states that have the death penalty differ from the number of homicides in states that do not have the death penalty.

SPSS QUESTIONS

Use the British Election Study data set entitled Version 2.2 Face-to-face Post-election Survey (with vote validation). It is dated May 2015 – September 2015.

www.britishelectionstudy.com/data-objects/cross-sectional-data/

Use this data set to create frequencies and crosstabs on select variables in order to make sense of this data set, which is very large.

1. Choose the following two variables: m02_1 *Politicians don't care what people like me think*; and, b02 *Which party did you vote for in the general election?* First, create a frequency table for each of them.

2. Next, run crosstabs on both variables.

3. Next, create a pivot table from the crosstabs results.

4. and 5. Finally, go to Graphs to create two cluster bars. You may follow the instructions in the textbook, or you may choose to use Graphs—Legacy Dialogs—Bar—Bar Charts—Cluster. Create the cluster bar both ways: with m02_1 on the *x*-axis and b02 on the *y*-axis, and then b02 on the *x*-axis and m02_1 on the *y*-axis.

JUST FOR FUN/CHALLENGE YOURSELF

1. After you create both cluster bars, give reasons why each one would be the best way to present the data.

ANSWER KEY

TRUE/FALSE QUESTIONS

1. True.

2. False. With the acquisition of new data, many data sets on the Internet are updated, so expect different numbers if you utilize the same link to data at a later time.

3. False. All the same procedures can also be used to make sense of small data sets.

4. False. Since there is only one variable at stake here (number of migrants settling in a new country), the more appropriate procedure would involve looking at frequencies.

5. True.

SPSS QUESTIONS

1. Frequency table for m02_1 *Politicians don't care what people like me think.*

Politicians don't care what people like me think.

		Frequency	Percent	Valid Percent	Cumulative Percent
Valid	Don't know	30	1.0	1.0	1.0
	Strongly disagree	89	3.0	3.0	4.0
	Disagree	762	25.5	25.5	29.5
	Neither agree nor disagree	578	19.4	19.4	48.8
	Agree	1029	34.4	34.4	83.3
	Strongly agree	499	16.7	16.7	100.0
	Total	2987	100.0	100.0	

Frequency table for b02 *Which party did you vote for in the general election?*

Which party did you vote for in the general election?

		Frequency	Percent	Valid Percent	Cumulative Percent
Valid	Refused	89	3.0	4.1	4.1
	Don't know	11	.4	.5	4.6
	Labour	677	22.7	30.8	35.4
	Conservatives	836	28.0	38.1	73.4
	Liberal Democrats	158	5.3	7.2	80.6
	Scottish National Party (SNP)	101	3.4	4.6	85.2
	Plaid Cymru	10	.3	.5	85.7
	Green Party	66	2.2	3.0	88.7
	United Kingdom Independence Party (UKIP)	234	7.8	10.7	99.3
	British National Party (BNP)	3	.1	.1	99.5
	Other	12	.4	.5	100.0
	Total	2197	73.6	100.0	
Missing	System	790	26.4		
Total		2987	100.0		

2. Crosstabs for both variables

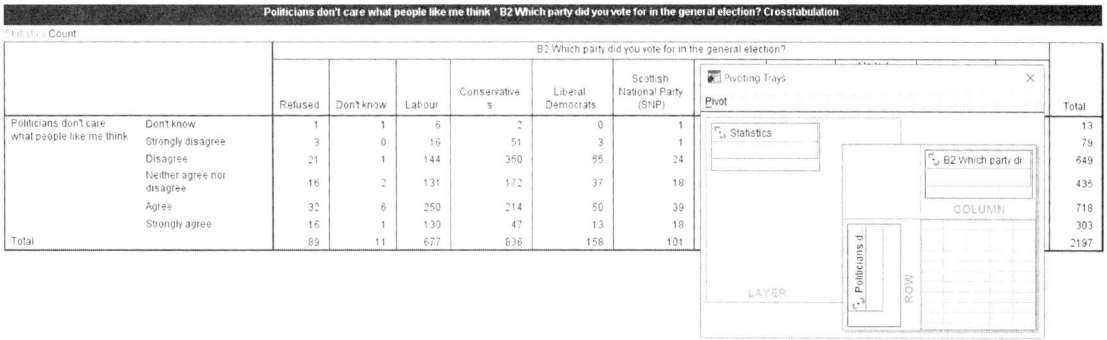

Politicians don't care what people like me think * B2 Which party did you vote for in the general election? Crosstabulation

Count

		B2 Which party did you vote for in the general election?											Total
		Refused	Don't know	Labour	Conservatives	Liberal Democrats	Scottish National Party (SNP)	Plaid Cymru	Green Party	United Kingdom Independence Party (UKIP)	British National Party (BNP)	Other	
Politicians don't care what people like me think	Don't know	1	1	6	2	0	1	0	0	2	0	0	13
	Strongly disagree	3	0	16	51	3	1	0	2	2	0	1	79
	Disagree	21	1	144	350	55	24	5	17	29	1	2	649
	Neither agree nor disagree	16	2	131	172	37	18	1	15	41	1	1	435
	Agree	32	6	250	214	50	39	4	24	94	0	5	718
	Strongly agree	16	1	130	47	13	18	0	8	66	1	3	303
Total		89	11	677	836	158	101	10	66	234	3	12	2197

3. Pivot table for both variables

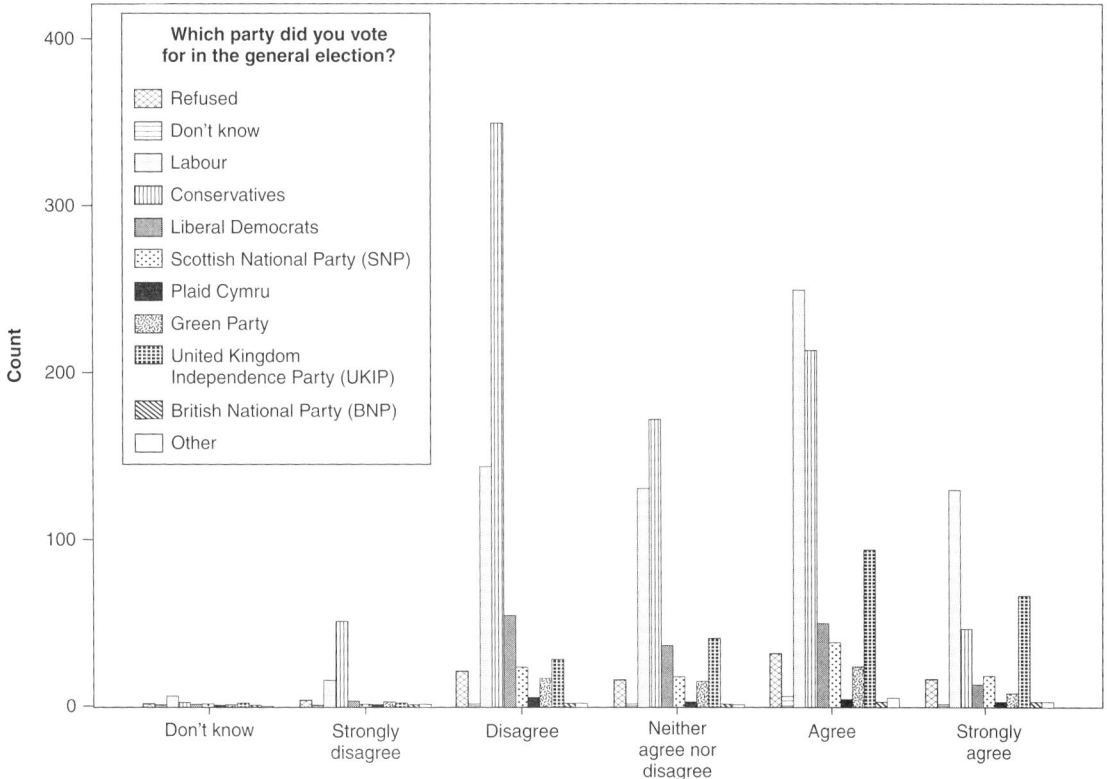

Politicians don't care what people like me think * B2 Which party did you vote for in the general election? Crosstabulation

Statistics Count

		B2 Which party did you vote for in the general election?						Total
		Refused	Don't know	Labour	Conservatives	Liberal Democrats	Scottish National Party (SNP)	
Politicians don't care what people like me think	Don't know	1	1	6	2	0	1	13
	Strongly disagree	3	0	16	51	3	1	79
	Disagree	21	1	144	350	55	24	649
	Neither agree nor disagree	16	2	131	172	37	18	435
	Agree	32	6	250	214	50	39	718
	Strongly agree	16	1	130	47	13	18	303
Total		89	11	677	836	158	101	2197

4. m02_1 *Politicians don't care what people like me think.*

Which party did you vote for in the general election?

- Refused
- Don't know
- Labour
- Conservatives
- Liberal Democrats
- Scottish National Party (SNP)
- Plaid Cymru
- Green Party
- United Kingdom Independence Party (UKIP)
- British National Party (BNP)
- Other

Politicians don't care what people like me think

b02 *Which party did you vote for in the general election?*

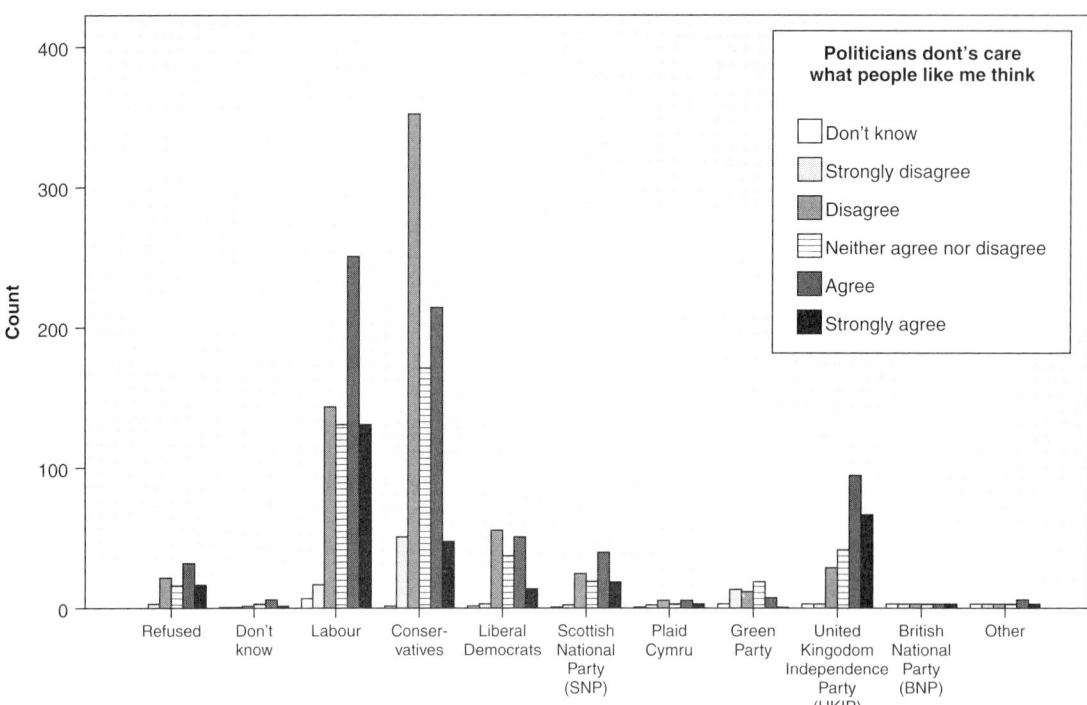

Which party did you vote for in the general...

JUST FOR FUN/CHALLENGE YOURSELF

1. When the cluster bar is produced with m02_1 *Politicians don't care what people like me think* on the x-axis, the distribution is based on the political party the respondents voted for. The focus is on the parties, so if a researcher wanted to know specific counts by party in order to hypothesize about them, then the first arrangement would be best. When the cluster bar is produced with b02 *Which party did you vote for in the general election* on the x-axis, the distribution is based on the respondents' thoughts of politicians not caring about what people like them think. The focus is on how the respondents ascribe to the idea of politicians who do not care about constituents' thoughts. If a researcher wanted to know specific counts by respondents' answers about politicians in order to hypothesize about them, then the second arrangement would be best.

Appendix A

Practice With Real Data!

Do you want more practice with data in psychology? Below are several studies with data collected by psychology research methods and statistics students at a Florida university. SPSS data sets are available for each study. You can find the data sets on the main website at **edge.sagepub.com/salkind6e**: click on **Resources for the Study Guide for Psychology** on the left-hand navigation.

    I.   Study One: Color Priming and Word Scramble

<u>Study Description:</u> In this study, researchers asked participants to complete a short word scramble task. Each participant received a list of 20 words (for example, BMTUH, HRCOP, LNATP, etc.) and they had to unscramble the word (THUMB, PORCH, and PLANT, respectively). The independent variable was the ink color in which the scrambled words were written: they were either in red ink, green ink, or black ink. Based on the theory that the color red impairs performance, we predicted that those unscrambling words written in red would unscramble fewer words in 3 minutes than participants given scrambled words in either black or green ink, which we did not expect to differ. Dependent variables included: number of anagrams solved (0 to 20), how challenging was the task (1 = not at all challenging to 0 = extremely challenging), how challenging do you think other participants found this task (1 = not at all challenging to 0 = extremely challenging), how frustrating was this task (1 = not at all frustrating to 9 = very frustrating), how do you think you did on the three-minute task (1 = very poorly to 9 = very well), and to what extent do you think the color of ink impacted your performance (1 = decreased the number I got correct to 9 = increased the number I got correct). We also asked participants to complete a nominal question: What color ink was used in the task (red, green, black, blue). Finally, participants provided demographic information (gender, age, ethnicity, whether English is the participant's primary language, whether the participant is a student, and whether the participant is color blind).

<u>Potential Analyses:</u> Using the data file, determine if we supported our hypotheses. You can run *t*-tests on all scaled variables (just select two colors from the manipulation), though I recommend running a simple one-way ANOVA on scaled variables. Use a factorial ANOVA by looking at color condition and gender! A chi-square would be appropriate for nominal dependent variables (like ink participants recall seeing as well as color-blind status of the participant). You can also run correlations for some of the dependent variables. The question asking about how challenging the task is for the participant may correlate highly with the question about how challenging the task is for others. Finally, find descriptive statistics for participant demographic characteristics. Keep in mind this data was collected in Miami, FL, so there are a lot of Hispanic participants!

SPSS FILE: Color Priming Survey

II. Study Two: Hindsight Bias

Study Description: In this study, we provided participants with the outcome from a rat study. That is, we told them the following in a questionnaire:

> "University researchers performed the following experiment. They injected blood from a mother rat into a female virgin rat immediately after the mother rat gave birth. After the injection, the virgin rat was placed in a cage with the newly born baby rats (after removing of their real mother). *Because of pregnancy hormones present in the blood of the mother rat, the researchers found that the virgin rat exhibited maternal behaviors towards the baby rats.*"

The above paragraph is our "hindsight maternal" condition. In our "hindsight nonmaternal" condition, we replaced the last sentence with this one: "*Despite pregnancy hormones present in the blood of the mother rat, the researchers found that the virgin rat did not exhibit maternal behaviors towards the baby rats.*" Finally, in our "foresight" condition, we provide participants with the following last sentence: "*There are several potential outcomes that may result from this experimental design: Given pregnancy hormones present in the blood of the mother rat, the virgin rat may either a) exhibit maternal behaviors toward the baby rats or b) fail to exhibit maternal behaviors toward the baby rats.*"

We predicted that those in the two hindsight conditions would find the outcome of the rat study obvious, even though these conditions predict different outcomes! That is, in hindsight the outcome looks obvious. The foresight condition provides a nice comparison group.

Our study here has one main independent variable with three levels (hindsight maternal, hindsight nonmaternal, and foresight). There are several dependent variables as well, including how surprised are you by the outcome (1 = not at all surprised to 9 = very surprised), how likely is it that researchers could replicate this finding (1 = very unlikely to 9 = very likely), and imagine the original researchers found the opposite result: how likely would they be to replicate this opposite result in the future (1 = very unlikely to 9 = very likely). We also asked a nominal dependent variable: is nature or nurture more controlling in this study. We then collected demographic information, including age, ethnicity, gender, and student status. For the two hindsight conditions only, we asked one final question: "We deceived you a bit: The original researchers did find the opposite outcome, not the one we told you they found. Does this real outcome surprise you (1 = not at all surprised to 9 = very surprised).

Potential Analyses: Using the data file, determine if we supported our hypotheses. You can run *t*-tests on all scaled variables (just select two of the three outcomes), though I recommend running a simple one-way ANOVA on scaled variables. Use a factorial ANOVA by looking at hindsight condition outcome and gender! A chi-square would be appropriate for nominal dependent variables (like the nature versus nurture variable). You can also run correlations for some of the dependent variables. The two likelihood of replication questions should negatively correlate with one another. Finally, find descriptive statistics for participant demographic characteristics.

SPSS FILE: Hindsight Bias Survey

III. Study Three: Belief Perseverance Survey

Study Description: In this study, we had participants read a description of a study in which some *original researchers* had children rank toys. The *researchers* focused on the second-favorite toy (the "forbidden toy," and either threatened the child to not play with it or simply set it aside. Later, the kids re-ranked the toys. In our own study, we told some participants that kids liked that second-ranked toy more when they were threatened to not play with it (our "More" condition). For others, we told them the kids ranked it lower (our "Less" condition). For remaining participants, we told them the child's preference for the forbidden toy did not change.

We asked participants if they were surprised by the outcome (1 = not at all surprised to 7 = very surprised), whether they think the results could be replicated elsewhere (1 = would not be replicated to 9 = would be replicated), and to decide whether the methods used were scientific (1 = not at all scientific to 7 = very scientific). These questions, however, were filler questions, since we told them that the original data was actually fake. We then asked them to imagine the results of the study were actually carried out. For follow-up questions, we asked them whether kids in the new study would find the forbidden toy preferable (−5 = less preferable to +5 = more preferable) and whether that preference rating would occur in a different location (also on a −5 to +5 scale). We asked whether failing to replicate the original (fake!) results would affect their trust in the study (1 = I'd find the new study questionable to 7 = I'd find the new study trustworthy). As a manipulation check, we asked them to recall the outcome of the original (fake!) study: the original kids found the forbidden toy more preferable, less preferable, or neither more nor less preferable. The study concluded with demographic questions (gender, age, ethnicity, student status, English as a first language).

We predicted that belief perseverance would lead participants to think the original study outcome would occur in the future study. That is, if told the original study led to more preference on the re-ranking, our participants would similarly think kids would prefer the toy more. If told less, then future kids would rank the toy in the real study as less. Finally, if told there was no change in preference, our participants will think future kids will similarly not change their preference.

Potential Analyses: Using the data file, determine if we supported our hypotheses. You can run *t*-tests on all scaled variables (just select two of the three outcomes: more, less, or neutral), though I recommend running a simple one-way ANOVA on scaled variables. Use a factorial ANOVA by looking at outcome condition and gender! A chi-square would be appropriate for nominal dependent variables (like whether they recall the original outcome of the study). You can also run correlations for some of the dependent variables. The replication questions should correlate with one another. Finally, find descriptive statistics for participant demographic characteristics.

SPSS FILE: Belief Perseverance Survey

IV. Study Four: Foot-in-the-door versus Door-in-the-face

Study Description: In this study, we wanted to see how willing participants were to participate in a second study that takes 30 minutes to complete after being told an original study takes either 5 minutes (Foot-in-the-door), 60 minutes (Door-in-the-face), or 30 minutes (Control) study.

In our "Foot-in-the-door" (FITD) condition, we asked some participants how willing they would be to participate in a study that takes 5 minutes (1 = not at all willing to 9 = very willing). We then asked if they were willing to participate right now (yes or no). Finally, we asked them how willing they would be to participate if the study took 30 minutes (1 = not at all willing to 9 = very willing).

In our "Door-in-the-face" (DITF) condition, we asked some participants how willing they would be to participate in a study that takes 60 minutes (1 = not at all willing to 9 = very willing). We then asked if they were willing to participate right now (yes or no). Finally, we asked them how willing they would be to participate if the study took 30 minutes (1 = not at all willing to 9 = very willing).

Finally, in our Control condition, we asked some participants how willing they would be to participate in a study that takes 30 minutes (1 = not at all willing to 9 = very willing). We then asked if they were willing to participate right now (yes or no).

We followed up each questionnaire by asking participants to provide their demographic information (gender, age, and ethnicity). Although we did not actually have them engage in a second study, we did ask them if we originally asked them to participate in a study that lasts 5 minutes, 30 minutes, or 60 minutes, which served as our manipulation check. We also asked about their initial impression of the first request they heard (1 = I did not want to participate to 9 = I wanted to participate).

<u>Predictions and Potential Analyses:</u> We predicted that those in the FITD condition would be more willing to participate in the first study (think it is 5 minutes) than both DITF condition (think it is 60 minutes) and Control condition (think it is 30 minutes). This first question, which is interval (1 to 9) can be analyzed with t-tests and ANOVAs. For the nominal second question, those in the FITD should say yes more often while those in the DITF and Control should say no more often (as seen with a chi-square). For the final question, only FITD and DITF are applicable (the question asked about a follow-up 30 minute study). Here, FITD and DITF should act similarly (the FITD participated once, so should do so again to be consistent; the DITF should say no to the first 60-minute study, so guilt might make them agree to the shorter 30-minute study). Note that you can reuse the Control first question for this last analysis as well as to compare all three levels of the independent variable in a one-way ANOVA. Correlations can be used to see if the two scaled willingness variables correlate.

SPSS FILE: Foot-In-The-Door Survey

Appendix B

Writing Up Your Results—Guidelines Based on APA Style

In a results section, your goal is to report the results of the data analyses used to test your hypotheses. To do this, you need to identify your data analysis technique, report your test statistic, and provide some interpretation of the results. Each analysis you run should be related to your hypotheses. If you analyze data that is exploratory, you need to indicate this. If your results are complicated—you have many conditions and/or many dependent measures—adding a table of figures can be helpful. See the APA Publication Manual for examples.

In reporting the results of statistical tests, report the descriptive statistics, such as means and standard deviations, as well as the test statistic, degrees of freedom, obtained value of the test, and the probability of the result occurring by chance (p value). Test statistics and p values should be rounded to two decimal places. All statistical symbols (sample statistics) that are not Greek letters should be italicized (M, SD, t, p, etc.).

When reporting a significant difference between two conditions, indicate the direction of this difference, i.e., which condition was more/less/higher/lower than the other condition(s). Assume that your audience has a professional knowledge of statistics. Do not explain how or why you used a certain test unless it is unusual (i.e., such as a nonparametric test). In the sections below, there are examples and tips for how to report your test statistics.

p VALUES

There are two ways to report p values. One way is to use the alpha level (the a priori criterion for the probability of falsely rejecting your null hypothesis), which is typically .05 or .01. Example: $F(1, 24) = 44.4, p < .01$. You may also report the exact p value (this is the preferred option if you want to make your data convenient for individuals conducting a meta-analysis on the topic). For example: $t(33) = 2.10, p = .03$. If your exact p value is less than .001, it is conventional to state merely $p < .001$. If you report exact p values, state early in the results section the alpha level used as a significance criterion for your tests. For example: "We used an alpha level of .05 for all statistical tests."

What if your results are in the predicted direction but not significant? If your p value is .10 or less, you can say your results were *marginally* significant. Example: Results indicated a marginally significant preference for pecan pie ($M = 3.45$, $SD = 1.11$) over cherry pie ($M = 3.00$, $SD = .80$), $t(5) = 1.25, p = .08$. If your p value is over .10, you can say your results revealed a nonsignificant trend in the predicted direction. Example: Results indicated a nonsignificant trend in the predicted direction indicating a preference for pecan pie ($M = 3.45$, $SD = 2.11$)

over cherry pie (M = 3.00, SD = 2.80), t(5) = 1.25, p = .26. Note: this is one example where disciplines vary but this is what is acceptable in social psychology.

DESCRIPTIVE STATISTICS

Mean and standard deviation are most clearly presented in parentheses:

- The sample as a whole was relatively young (M = 19.22, SD = 3.45).

- The average age of students was 19.22 years (SD = 3.45).

Percentages are also most clearly displayed in parentheses with no decimal places:

- Nearly half (49%) of the sample was married.

t-TESTS

There are several different research designs that utilize a t-test for the statistical inference testing. The differences between one-sample t-tests, related measures t-tests, and independent-samples t-tests are so clear to the knowledgeable reader that most journal editors eliminate the elaboration of which type of t-test has been used. Additionally, the descriptive statistics provided will identify further which variation was employed. It is important to note that we assume that all p values represent two-tailed tests unless otherwise noted and that independent-samples t-tests use the pooled variance approach (based on an equal variances assumption) unless otherwise noted:

- There was a significant effect for gender, t(54) = 5.43, p < .001, with men (M = 3.45, SD = 2.3) receiving higher scores than women (M = 1.22, SD = .98).

- Results indicate a significant preference for pecan pie (M = 3.45, SD = 1.11) over cherry pie (M = 3.00, SD = .80), t(15) = 4.00, p = .001.

- Non-student participants had a mean age of 27.4 (SD = 12.6), which was significantly older than the 21.2 years (SD = 8.21) of the university students, t(35) = 2.95, p = 0.01.

- Students taking statistics courses in psychology at the University of Washington reported studying more hours for tests (M = 121, SD = 14.2) than did UW college students in general (M = 85, SD = 12.2), t(33) = 2.10, p = .034.

- The 25 participants had an average difference from pretest to posttest anxiety scores of −4.8 (SD = 5.5), indicating the anxiety treatment resulted in a significant decrease in anxiety levels, t(24) = −4.36, p = .005 (one-tailed).

- The 36 participants in the treatment group (M = 14.8, SD = 2.0) and the 25 participants in the control group (M = 16.6, SD = 2.5) demonstrated a significance difference in performance (t[59] = −3.12, p = .01); as expected, the visual priming treatment inhibited performance on the phoneme recognition task.

- UW students taking statistics courses in psychology had higher IQ scores (M = 121, SD = 14.2) than did those taking statistics courses in statistics (M = 117, SD = 10.3), t(44) = 1.23, p = .09.

- Over a two-day period, participants drank significantly fewer drinks in the experimental group (M = 0.667, SD = 1.15) than did those in the wait-list control group (M = 8.00, SD = 2.00), t(4) = −5.51, p = .005.

ANOVA AND POST HOC TESTS

ANOVAs are reported like the t-test, but there are two degrees-of-freedom numbers to report. First report the between-groups degrees of freedom, then report the within-groups degrees of freedom (separated by a comma). After that report the F statistic (rounded off to two decimal places) and the significance level.

One-way ANOVA

- The 12 participants in the high dosage group had an average reaction time of 12.3 seconds ($SD = 4.1$); the 9 participants in the moderate dosage group had an average reaction time of 7.4 seconds ($SD = 2.3$), and the 8 participants in the control group had a mean of 6.6 ($SD = 3.1$). The effect of dosage, therefore, was significant, $F(2,26) = 8.76$, $p = .012$.

- A one-way analysis of variance showed that the effect of noise was significant, $F(3,27) = 5.94$, $p = .007$. Post hoc analyses using the Scheffé post hoc criterion for significance indicated that the average number of errors was significantly lower in the white noise condition ($M = 12.4$, $SD = 2.26$) than in the other two noise conditions (traffic and industrial) combined ($M = 13.62$, $SD = 5.56$), $F(3, 27) = 7.77$, $p = .042$.

- Tests of the four a priori hypotheses were conducted using Bonferroni adjusted alpha levels of .0125 per test (.05/4). Results indicated that the average number of errors was significantly lower in the silence condition ($M = 8.11$, $SD = 4.32$) than were those in both the white noise condition ($M = 12.4$, $SD = 2.26$), $F(1, 27) = 8.90$, $p = .011$, and in the industrial noise condition ($M = 15.28$, $SD = 3.30$), $F(1, 27) = 10.22$, $p = .007$. The pairwise comparison of the traffic noise condition with the silence condition was nonsignificant. The average number of errors in all noise conditions combined ($M = 15.2$, $SD = 6.32$) was significantly higher than those in the silence condition ($M = 8.11$, $SD = 3.30$), $F(1, 27) = 8.66$, $p = .009$.

Multiple Group/Factorial (Independent Variable) ANOVA

- There was a significant main effect for treatment, $F(1, 145) = 5.43$, $p < .01$, and a significant interaction, $F(2, 145) = 3.13$, $p < .05$.

- The cell sizes, means, and standard deviations for the 3 X 4 factorial design are presented in Table 1. The main effect of dosage was marginally significant ($F[2,17] = 3.23$, $p = .067$), as was the main effect of diagnosis category, $F(3,17) = 2.87$, $p = .097$. The interaction of dosage and diagnosis, however, has significant, $F(6,17) = 14.2$, $p = .0005$.

- Attitude change scores were subjected to a two-way analysis of variance having two levels of message discrepancy (small, large) and two levels of source expertise (high, low). All effects were statistically significant at the .05 significance level. The main effect of message discrepancy yielded an F ratio of $F(1, 24) = 44.4$, $p < .001$, indicating that the mean change score was significantly greater for large-discrepancy messages ($M = 4.78$, $SD = 1.99$) than for small-discrepancy messages ($M = 2.17$, $SD = 1.25$). The main effect of source expertise yielded an F ratio of $F(1, 24) = 25.4$, $p < .01$, indicating that the mean change score was significantly higher in the high-expertise message source ($M = 5.49$, $SD = 2.25$) than in the low-expertise message source ($M = 0.88$, $SD = 1.21$). The interaction effect was nonsignificant, $F(1, 24) = 1.22$, $p > .05$.

- A two-way analysis of variance yielded a main effect for the diner's gender, $F(1,108) = 3.93$, $p < .05$, such that the average tip was significantly higher for men ($M = 15.3\%$, $SD = 4.44$) than for women ($M = 12.6\%$, $SD = 6.18$). The main effect of touch was nonsignificant, $F(1, 108) = 2.24$, $p > .05$. However, the interaction effect was significant, $F(1, 108) = 5.55$, $p < .05$, indicating that the gender effect was greater in the touch condition than in the non-touch condition.

CHI-SQUARE

Chi-square statistics are reported with degrees of freedom and sample size in parentheses, the Pearson chi-square value (rounded to two decimal places), and the significance level:

- The percentage of participants that were married did not differ by gender, $\chi^2(1, N = 90) = 0.89$, $p > .05$.

- The sample included 30 respondents who had never married, 54 who were married, 26 who reported being separated or divorced, and 16 who were widowed. These frequencies were significantly different, $\chi^2 (3, N = 126) = 10.1$, $p = .017$.

- As can be seen by the frequencies cross tabulated in Table XX, there is a significant relationship between marital status and depression, $\chi^2 (3, N = 126) = 24.7$, $p < .001$.

- The relation between these variables was significant, $\chi^2 (2, N = 170) = 14.14$, $p < .01$. Catholic teens were less likely to show an interest in attending college than were Protestant teens.

- Preference for the three sodas was not equally distributed in the population, $\chi^2 (2, N = 55) = 4.53$, $p < .05$.

CORRELATIONS

Correlations are reported with the degrees of freedom (which is N-2) in parentheses and the significance level:

- The two variables were strongly correlated, $r(55) = .49$, $p < .01$.